GUIDE TO MITIGATING SPACECRAFT CHARGING EFFECTS

JPL SPACE SCIENCE AND TECHNOLOGY SERIES

Joseph H. Yuen, Editor-in-Chief

Published Titles in This Series

Fundamentals of Electric Propulsion: Ion and Hall Thrusters
Dan M. Goebel and Ira Katz

Synthetic Aperture Radar Polarimetry
Jacob Van Zyl and Yunjin Kim

Guide to Mitigating Spacecraft Charging Effects
Henry B. Garrett and Albert C. Whittlesey

GUIDE TO MITIGATING SPACECRAFT CHARGING EFFECTS

Henry B. Garrett
Albert C. Whittlesey
Jet Propulsion Laboratory
California Institute of Technology

WILEY

A JOHN WILEY & SONS, INC., PUBLICATION

Published by John Wiley & Sons, Inc., Hoboken, New Jersey.
Published simultaneously in Canada.

Library of Congress Cataloging-in-Publication Data:

Garrett, Henry B.
Guide to mitigating spacecraft charging effects / Henry B. Garrett, Albert C. Whittlesey. — 1st ed.
p. cm. — (JPL space science and technology series)
Includes bibliographical references and index.
ISBN 978-1-118-18645-9 (hardback)
1. Space vehicles–Electrostatic charging. 2. Electric discharges–Prevention. I. Title.
TL1492.G37 2012
629.47–dc23

2011036330

10 9 8 7 6 5 4 3 2 1

CONTENTS

NOTE FROM THE SERIES EDITOR

The Jet Propulsion Laboratory (JPL) Space Science and Technology Series broadens the range of the ongoing JPL Deep Space Communications and Navigation Series to include disciplines other than communications and navigation in which JPL has made important contributions. The books are authored by scientists and engineers with many years of experience in their respective fields, and lay a foundation for innovation by communicating state-of-the-art knowledge in key technologies. The series also captures fundamental principles and practices developed during decades of space exploration at JPL, and it celebrates the successes achieved. These books will serve to guide a new generation of scientists and engineers.

We would like to thank the Office of the Chief Scientist and Chief Technologist for their encouragement and support. In particular, we would like to acknowledge the support of Thomas A. Prince, former JPL Chief Scientist; Erik K. Antonsson, former JPL Chief Technologist; Daniel J. McCleese, JPL Chief Scientist; and Paul E. Dimotakis, JPL Chief Technologist.

JOSEPH H. YUEN, Editor-in-Chief

JPL Space Science and Technology Series
Jet Propulsion Laboratory
California Institute of Technology

FOREWORD

I am very pleased to commend the Jet Propulsion Laboratory (JPL) Space Science and Technology Series, and to congratulate and thank the authors for contributing their time to these publications. It is always difficult for busy scientists and engineers, who face the constant pressures of launch dates and deadlines, to find the time to tell others clearly and in detail how they solved important and difficult problems, so I applaud the authors of this series for the time and care they devoted to documenting their contributions to the adventure of space exploration.

JPL has been NASA's primary center for robotic planetary and deep-space exploration since the Laboratory launched the nation's first satellite, *Explorer 1*, in 1958. In the 50 years since this first success, JPL has sent spacecraft to all the planets except Pluto, studied our own planet in wavelengths from radar to visible, and observed the universe from radio to cosmic ray frequencies. Current plans call for even more exciting missions over the next decades in all these planetary and astronomical studies, and these future missions must be enabled by advanced technology that will be reported in this series. The JPL Deep Space Communications and Navigation book series captured the fundamentals and accomplishments of these two related disciplines, and we hope that this new series will expand the scope of those earlier publications to include other space science, engineering, and technology fields in which JPL has made important contributions.

I look forward to seeing many important achievements captured in these books.

CHARLES ELACHI, Director

Jet Propulsion Laboratory
California Institute of Technology

PREFACE

The purpose of this book is multifold. First, it serves as a single reference source that contains suggested detailed spacecraft design requirements and procedures to minimize the effects of spacecraft charging and to limit the effects of the resulting electrostatic discharge. Second, it contains supplementary material and references to aid in understanding and assessing the magnitude of the phenomenon. The book is intended to describe conditions under which spacecraft charging might be an issue, generally explain why the problem exists, list typical design solutions, and provide an introduction to the process by which design specifics should be resolved. The document is also intended to be an engineering tool, and is written at the graduate engineering level for use by aerospace engineers, system designers, program managers, and others concerned with space environment effects on spacecraft. It is not possible to place all the necessary knowledge into one document to be used as a cookbook; therefore, this document should be used as a preliminary reference and/or checklist only, primarily to identify if spacecraft charging is an issue for a particular mission. Once that determination has been made, it is recommended that project managers employ experienced electrostatic discharge (ESD)- and plasma-interactions engineers and scientists to perform detailed mission and spacecraft design analyses. Much of the environmental data and material response information has been adapted from published and unpublished scientific literature for use in this document. In particular, this book is intended as the textbook form of its source, NASA Technical Handbook NASA-HDBK-4002A, March 3, 2011.

Spacecraft charging, defined as the buildup of charge in and on spacecraft materials, is a significant phenomenon for spacecraft in certain Earth and other planetary environments. Design for control and mitigation of surface charging, the buildup of charge on the exterior surfaces of a spacecraft related to space plasmas, was treated in detail in NASA TP-2361, *Design Guidelines for Assessing and Controlling Spacecraft Charging Effects* (1984). Design for control and mitigation of internal charging, the buildup of charge on the interior parts of a spacecraft from higher-energy particles, was treated in detail in the original version of NASA-HDBK-4002, *Avoiding Problems Caused by Spacecraft On-Orbit Internal Charging Effects* (1999). NASA-HDBK-4002 was written as a companion document to NASA TP-2361.

Since the original writing of the two documents there have been developments in the understanding of spacecraft charging issues and mitigation solutions as well as advanced technologies needing new mitigation solutions. That, and the desire to merge the two documents, was the motivation for this revision. As in the heritage documents, the story still has unfinished business, and the proper way to address design issues for a specific satellite is to have skilled ESD-knowledgeable engineers as part of the design team for those programs and missions where space charging is an issue.

Acknowledgments

This document would have benefited substantially if our friend A. Robb Frederickson (July 5, 1941–April 5, 2004) had not died at a relatively early age. Robb worked for many years as part of Air Force Research Laboratories and, during his final years, as part of JPL. We would like to acknowledge his many valuable insights and comments that inspired this rewrite, as well as his technical contributions to the field of spacecraft charging. We also owe special thanks to N. John Stevens (July 11, 1930–March 11, 2010), who worked at NASA Lewis Research Center and at TRW. John was a great friend and source of inspiration for this book and its predecessors.

The discipline of spacecraft charging has had many practitioners since at least 1970, and this document benefits by the research performed and publications presented by those persons, as well as personal conversations with many of them. These contributors, we hope, have been properly referenced for their specific contributions within this document. We would also like to thank the committee of experts who reviewed the document before publication. We would also like to acknowledge several people who helped prepare preliminary drafts, read earlier drafts, and provided much appreciated feedback on the document. Randy Swimm prepared the initial combined draft. Major reviewers included Victoria Davis, Barbara Gardner, and Dale Ferguson. Our colleague Ira Katz was a sounding board for the contents as the document evolved.

HENRY B. GARRETT
ALBERT C. WHITTLESEY

June 2011

1 Introduction

This book documents engineering guidelines and design practices that can be used by spacecraft designers to minimize the detrimental effects of spacecraft surface and internal charging in certain space environments. Chapter 2 covers space charging/electrostatic discharge background and orientation; Chapter 3, design guidelines; Chapter 4, spacecraft test techniques; Chapter 5, control and monitoring methods; and Chapter 6, materials that should or should not be considered for charging control. The appendixes contain a collection of useful material intended to support the main body of the document. Despite our desire that this be an all-encompassing guideline, this document cannot do that. It is a narrowly focused snapshot of existing technology, not a research report, and does not include certain related technologies or activities as clarified further below.

In-space charging effects are caused by interactions between the in-flight plasma environment and spacecraft materials and electronic subsystems. Possible detrimental effects of spacecraft charging include disruption of or damage to subsystems (such as power, navigation, communications, or instrumentation) because of field buildup and electrostatic discharge as a result of a spacecraft's passage through the space plasma and high-energy particle environments. Charges can also attract contaminants, affecting thermal properties, optical instruments, and solar arrays; and they can change particle trajectories, thus affecting plasma-measuring instruments. NASA RP-1375, *Failures and Anomalies Attributed to Spacecraft Charging* [1], lists and describes some spaceflight failures caused by inadequate designs.

This book applies to Earth-orbiting spacecraft that pass through the hazardous regions identified in Figs. 1-1 and 1-2 [medium Earth orbit (MEO), low Earth orbit (LEO), and geosynchronous Earth orbit (GEO), with less focus on polar Earth orbit (PEO)], as well as spacecraft in other energetic plasma environments, such as those at Jupiter and Saturn, and interplanetary solar wind charging environments. Designs for spacecraft with orbits in these regions should be evaluated for the threat of external (surface) and/or internal charging, as noted. NASA RP-1354, *Spacecraft Environments Interactions: Protecting Against the Effects of Spacecraft Charging* [2], describes environmental interaction mitigation design techniques at an introductory level.

Guide to Mitigating Spacecraft Charging Effects, First Edition. Henry B. Garrett, Albert C. Whittlesey.

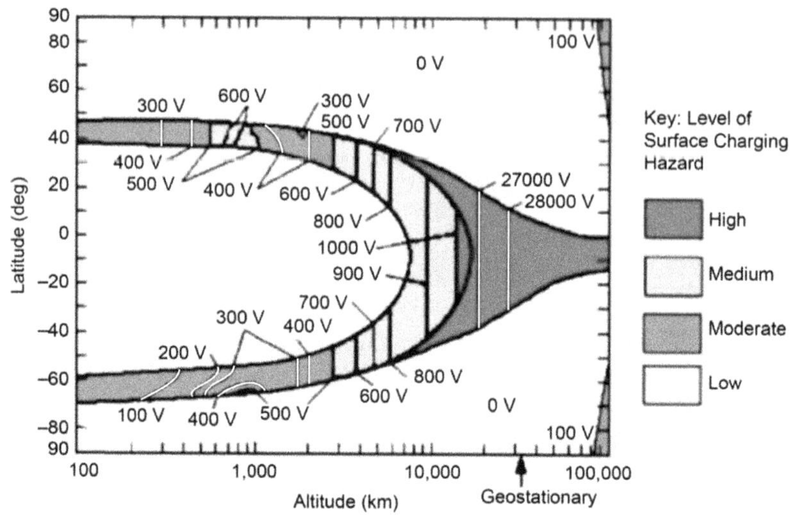

Fig. 1-1. Earth regimes of concern for on-orbit surface charging hazards for spacecraft passing through the latitude and altitude indicated. See Whittlesey et al. [9] for an alternative reference with the "Wishbone" chart. (From [8].) (*See insert for color representation of the figure*.)

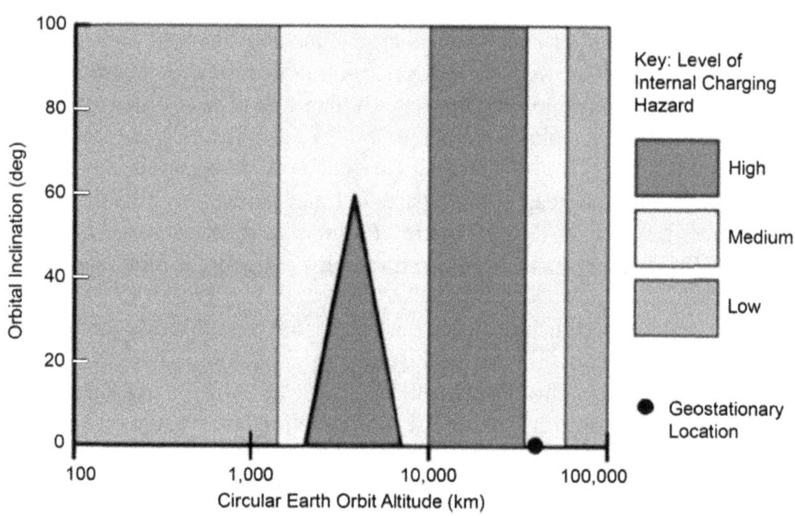

Fig. 1-2. Earth regimes of concern for on-orbit internal charging hazards for spacecraft with circular orbits. (*See insert for color representation of the figure*.)

Specifically, this book does not address LEO spacecraft charging at orbital inclinations such that the auroral zones are seldom encountered. That region is the purview of NASA-STD-4005 [3] and NASA-HDBK-4006 [4]. The book is intended to be complementary to those standards and applies to other regions.

In particular, mitigation techniques for low-inclination LEO orbits may differ from those that apply to regions covered by this book. Spacecraft in orbits, such as GEO transfer orbits that spend time in both regimes, should use mitigation techniques that apply to both regimes. It also does not include such topics as the following:

- Landed assets (e.g., lunar or Martian landers) and their electrostatic dust charging
- Spacecraft sources of charging (such as various types of electric propulsion or plasma sources)
- International Space Station (ISS)–specific design considerations (these encompass substantially different design concerns that are unique to the ISS)
- Solar-array-driven charging (see references [3,4])
- Magnetic field interactions relating to spacecraft charging (refer to tether and ISS sources for information)
- Mars-, Venus-, asteroid-, or Moon-specific charging environments (including surface charging environments)
- Plasma contactors in detail (see ISS references)
- Extravehicular activity needs (see ISS references)
- Specific design advice for pending or future projects
- Highly elliptical (Molniya) orbits

Figures 1-1 and 1-2 illustrate the approximate regions of concern for charging as defined in this book. Figure 1-1 is to be interpreted as the worst-case surface charging that may occur in the near-Earth environment. The north/south latitudinal asymmetry assumes that the magnetic North Pole is tilted as much as possible for this view. Potentials are calculated for an aluminum sphere in shadow. Note that at altitudes above 400 km, spacecraft charging can exceed 400 to 500 V, which has the possibility of generating discharges. Indeed, the Defense Meteorological Satellite Program (DMSP) and other satellites have reported significant charging in the auroral zones many times (as high as -4000 V), and one satellite [Advanced Earth Observation Satellite II (ADEOS-II)] at 800 km experienced total failure due to spacecraft charging [5–7].

Figure 1-2, which illustrates Earth's internal charging threat regions, is estimated assuming averages over several orbits since the internal charging threat usually has a longer time scale and reflects the approximate internal charging threat for satellites with the indicated orbital parameters. It is intended to illustrate the approximate regions of concern for internal electrostatic discharge (IESD).

In this book, the distinction between surface charging and internal charging is that *internal charging* is caused by energetic particles that can penetrate and deposit charge very close to a victim site. *Surface charging* occurs on areas that can be seen and touched on the outside of a spacecraft. Surface discharges occur on or near the outer surface of a spacecraft, and discharges must be coupled to an interior affected site rather than directly to the victim. Energy from surface arcs is attenuated by the coupling factors necessary to get to victims (most often inside the spacecraft) and therefore is less of a threat to electronics. External wiring and antenna feeds, of course, are susceptible to this threat. Internal charging, by contrast, may cause a discharge directly to a victim pin or wire with very little attenuation if caused by electron deposition in circuit boards, wire insulation, or connector potting.

Geosynchronous orbit (a circular orbit in the equatorial plane of Earth at about 35,786 km altitude) is perhaps the most common example of a region where spacecraft are affected by spacecraft charging, but the same problem can occur at lower Earth altitudes, in Earth polar orbits, at Jupiter, and at other places where spacecraft can fly. Internal charging is sometimes called deep dielectric charging or buried charging. Use of the word *dielectric* can be misleading, since ungrounded internal conductors can also present an internal electrostatic discharging threat to spacecraft. This book details the methods necessary to mitigate both in-flight surface and internal charging concerns as the physics and design solutions for both are often similar.

REFERENCES

[1] R. D. Leach and M. B. Alexander, Eds., *Failures and Anomalies Attributed to Spacecraft Charging*, NASA Reference Publication 1375, National Aeronautics and Space Administration, August 1995. This document has a very good list of specific space incidents that have been attributed to electrostatic discharges in space. It does not discriminate between surface charging or internal charging, but that is usually difficult to determine or does not appear in public literature.

[2] J. L. Herr and M. B. McCollum, *Spacecraft Environments Interactions: Protecting Against the Effects of Spacecraft Charging*, NASA-RP-1354, National Aeronautics and Space Administration, 1994.

[3] D. C. Ferguson, *Low Earth Orbit Spacecraft Charging Design Standard*, NASA-STD-4005, 16 pages, National Aeronautics and Space Administration, June 3, 2007.

[4] D. C. Ferguson, *Low Earth Orbit Spacecraft Charging Design Handbook*, NASA-HDBK-4006, 63 pages, National Aeronautics and Space Administration, June 3, 2007.

[5] D. L. Cooke, "Simulation of an Auroral Charging Anomaly on the DMSP Satellite," *36th Aerospace Sciences Meeting and Exhibit*, Reno, Nevada, AIAA-98-0385, January 12–15, 1998.

[6] S. Kawakita, H. Kusawake, M. Takahashi, H. Maejima, J. Kim, S. Hosoda, M. Cho, K. Toyoda, and Y. Nozaki, "Sustained Arc Between Primary Power Cables of a Satellite," *2nd International Energy Conversion Engineering Conference*, Providence, Rhode Island, August 16–19, 2004. Contains description of ADEOS-II satellite failure analysis. See also Maejima et al. [7].

[7] H. Maejima, S. Kawakita, H. Kusawake, M. Takahashi, T. Goka, T. Kurosaki, M. Nakamura, K. Toyoda, and M. Cho, "Investigation of Power System Failure of a LEO

Satellite," *2nd International Energy Conversion Engineering Conference*, Providence, Rhode Island, August 16–19, 2004. Contains description of ADEOS-II satellite failure analysis.

[8] R. W. Evans, H. B. Garrett, S. Gabriel, and A. C. Whittlesey, "A Preliminary Spacecraft Charging Map for the Near Earth Environment," *Spacecraft Charging Technology Conference*, Naval Postgraduate School, Monterey, California, November 1989. This original reference paper was omitted from the conference proceedings. See Whittlesey et al. [9] for an alternative reference with the "wishbone" chart.

[9] A. Whittlesey, H. B. Garrett, and P. A. Robinson, Jr., "The Satellite Space Charging Phenomenon, and Design and Test Considerations," *IEEE International EMC Symposium*. Anaheim, California, 1992.

2 Introduction to the Physics of Charging and Discharging

The fundamental physical concepts that account for space charging are described in this chapter. The appendices expand this description by means of equations and examples.

2.1 PHYSICAL CONCEPTS

Spacecraft charging occurs when charged particles from the surrounding plasma and energetic particle environment stop on the spacecraft: either on the surface, on interior parts, in dielectrics, or in conductors. Other items affecting charging include biased solar arrays or plasma emitters. Charging can also occur when photoemission occurs; that is, solar photons cause surfaces to emit photoelectrons. Events after that determine whether or not the charging causes problems.

2.1.1 Plasma

A *plasma* is a partially ionized gas in which some of the atoms and molecules that make up the gas have some or all of their electrons stripped off, leaving a mixture of ions and electrons that can develop a sheath that can extend over several Debye lengths. Except for LEO, where ionized oxygen (O^+) is the most abundant species, the simplest ion, a proton (corresponding to ionized hydrogen, H^+), is generally the most abundant ion in the environments considered here. The energy of the plasma, its electrons and ions, is often described in units of electron volts (eV). This is the kinetic energy that is given to the electron or ion if it is accelerated by an electric potential of that many volts. Whereas temperature (T) is generally used to describe the disordered microscopic motion of a group of particles, plasma physicists also use it as another unit of measure to describe the kinetic energy of the plasma. For electrons, numerically T(K) equals T(eV) $\times$ 11,604; that is, 4300 eV is equivalent to 50 million degrees kelvin (K).

The kinetic energy of a particle is given by the equation

$$E = \tfrac{1}{2}mv^2 \qquad (2.1\text{-}1)$$

Guide to Mitigating Spacecraft Charging Effects, First Edition. Henry B. Garrett, Albert C. Whittlesey.

where

E = energy
m = mass of the particle
v = velocity of the particle

Because of the difference in mass ($\sim$1 : 1836 for electrons to protons), electrons in a plasma in thermal equilibrium generally have a velocity about 43 times that of protons. This translates into a net instantaneous flux or current of electrons onto a spacecraft that is much higher than that of the ions [typically, nanoamperes per square centimeter (nA/cm^2) for electrons versus picoamperes per square centimeter (pA/cm^2) for protons at geosynchronous orbit]. This difference in flux is one reason for the charging effects observed (a surplus of negative charges on affected regions). For electrons, numerically the velocity (v_e) equals $\sqrt{E} \times 593$ km/s and for protons the velocity (v_p) equals $\sqrt{E} \times 13.8$ km/s, when E is in eV.

Although a plasma may be described by its average energy, there is actually a distribution of energies. The rate of charging in the interior of the spacecraft is a function of the flux versus energy, or spectrum, of the plasma at energies well in excess of the mean plasma energies [for GEO, the mean plasma energy may reach a few tens of kiloelectron volts (keV)]. Surface charging is usually correlated with electrons in the approximate 0 to 50 keV energy range, while significant internal charging is associated with the high-energy electrons [100 keV to 3 megaelectron volts (MeV)].

A simple plasma and its interactions with a surface are illustrated in Figs. 2-1 and 2-2. The electrons (e$^-$) and ions (represented by H$^+$ in Fig. 2-2) are moving in random directions (omnidirectional) and with different speeds (a spectrum of energies). Figure 2-2 illustrates surface charging. (Exterior surfaces are shown; the interior is similar.) To estimate surface charging, both the electron and ion

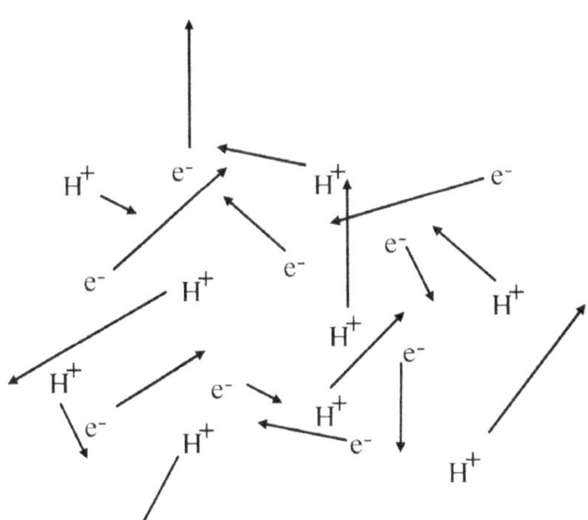

Fig. 2-1. Simple plasma.

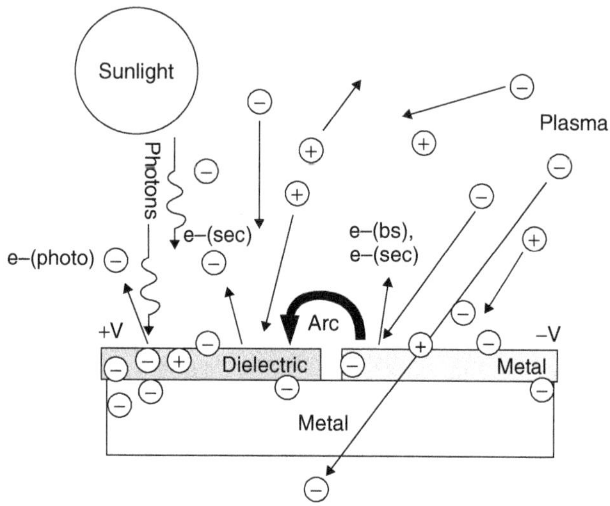

Fig. 2-2. Plasma interactions with spacecraft surfaces.

spectra should be known from about 1 eV to 100 keV. Although fluxes might be directed, omnidirectional fluxes are assumed in this document because spacecraft orientation relative to the plasma is often not well defined.

2.1.2 Penetration

Electrons and ions will penetrate matter. The depth of penetration of a given species (electron, proton, or other ion) depends on its energy, its atomic mass, and the composition of the target material. Figure 2-3 shows the mean penetration range versus the energy of electrons and protons into aluminum and represents the approximate penetration depth into a slab of aluminum. To first order, only particles with an energy corresponding to a range greater than the spacecraft shield thickness can penetrate the spacecraft interior. If the material is not aluminum, an equivalent penetration depth is roughly the same number of grams per square centimeter of the material's thickness.

In this document we use the terms *surface charging* and *internal charging*. In the literature the terms *buried dielectric charge* or *deep dielectric charge* for internal charging are also used, but these terms are misleading because they give the impression that only dielectrics can accumulate charge. Although dielectrics can accumulate charge and discharge to cause damage, ungrounded conductors can also accumulate charge and must also be considered an internal charging threat. In fact, ungrounded conductors can discharge with a higher peak current and a higher rate of change of current than a dielectric and can be a greater threat.

Based on typical spacecraft construction, there is usually an interior section, referred to in this document as internal. It is assumed that this interior section has shielding of at least 3 mils of aluminum equivalent, corresponding to electron

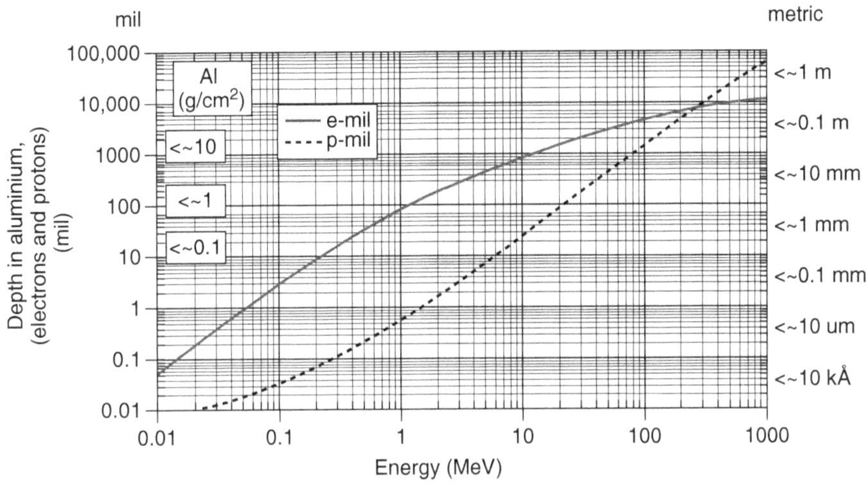

Notes:

1. Protons stop close to the mean depth shown, while electrons are deposited in a larger range around the given depth.
2. Surface charging: about 0 to 50 keV electrons.
3. Transition between surface and internal charging: about 50 to 100 keV.
4. Internal charging: greater than 100 keV.
5. For GEO orbits, the practical range of interest for internal charging is 0.1 to 3 MeV (~3 to 110 mils of aluminum thickness).
6. Data for figure from ESTAR and PSTAR, at http://physics.nist.gov/Star. [18]. (Note: The Web databases ESTAR, PSTAR, and ASTAR are used to calculate stopping powers, ranges, and related quantities for electrons, protons, and helium ions.)

Fig. 2-3. Electron/proton mean penetration energy ranges in aluminum.

energies greater than 0.1 MeV. Surface charging would be the outer layers of the spacecraft, corresponding to 2 mils of aluminum or 0 to 50 keV electrons. Obviously, the surface/internal charging cutoff depends on spacecraft construction. Protons are often not considered for spacecraft charging because the greater impinging flux of electrons at the same energy and (for internal charging) the lesser penetration of protons reduces the internal flux to a negligible amount. Higher atomic mass particles are even less of a threat because of their much lower fluxes.

Because electrons may stop at a depth that is less than their maximum penetration depth and because the electron spectrum is continuous, the penetration depth/charging region will be continuous, ranging from the charges deposited on the exterior surface to those deposited deep in the interior. Internal charging as used here often is equivalent to "inside the Faraday cage." For a spacecraft that is built with a Faraday cage thickness of 30 or more mils of aluminum equivalent, this would mean that internal effects deal with the portion of the electron spectrum above 500 keV and the proton spectrum above 10 MeV. At GEO orbits, the practical range of energy for internal charging is 100 keV to about 3 MeV,

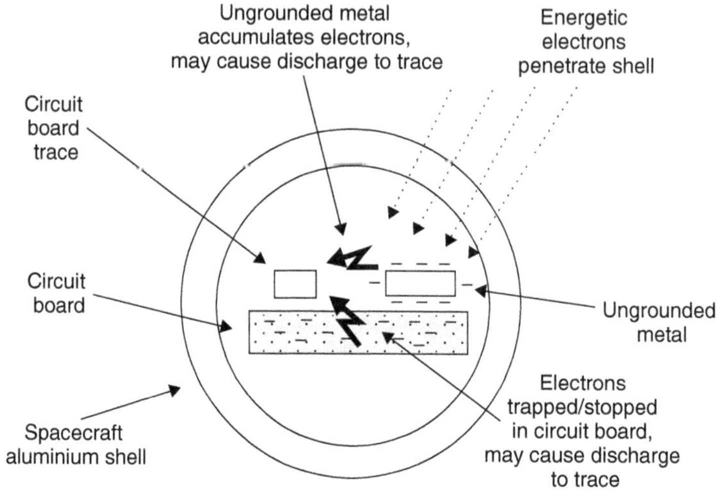

Fig. 2-4. Illustration of internal charging.

bounded on the lower end by the fact that most spacecraft have at least 3 mils of shielding and on the upper end by the fact that, as will be shown later, common GEO environments above 3 MeV do not have enough plasma flux to cause internal charging problems.

Figure 2-4 illustrates the concept that energetic electrons will penetrate interior portions of a spacecraft. Having penetrated, the electrons may be stopped in dielectrics or on ungrounded conductors. If too many electrons accumulate, the resulting high electric fields inside the spacecraft may cause an electrostatic discharge (ESD) to a nearby victim circuit. Note that the internal charging resembles surface charging, with the exception that circuits are rarely exposed victims on the exterior surface of a spacecraft, and thus (with the condition that charging rates are slower) internal charging results in a greater direct threat to circuits.

The term *ESD* in this document is general or may refer to surface discharges. The term *internal ESD* (IESD) refers to ESDs on the interior regions of a spacecraft as defined above.

2.1.3 Charge Deposition

The first step in analyzing a design for the internal charging threat is to determine the charge deposition inside the spacecraft. It is important to know the amount of charge deposited in or on a given material, as well as the deposition rate, as these determine the distribution of the charge and hence the local electric fields. An electrical breakdown (discharge) will occur when the local electric field exceeds the dielectric strength of the material or between dissimilar surfaces with a critical potential difference. The actual breakdown can be triggered by a variety of mechanisms including the plasma cloud associated with a micrometeoroid or space debris impact. The amplitude and duration of the resulting pulse are dependent on the charge deposited. These values in turn determine how much damage may be done to spacecraft circuitry.

Charge deposition is a function not only of the spacecraft configuration but also of the external electron spectrum. Given an electron spectrum and an estimate of the exterior shielding, the penetration depth versus the energy chart (Fig. 2-3) permits an estimate of electron deposition as a function of depth for any given equivalent thickness of aluminum, from which the likelihood of a discharge can be predicted. Because of such complexities as hardware geometries, however, it is normally better to run an electron penetration or radiation shielding code to more accurately determine the charge deposited at a given material element within a spacecraft. Appendixes B and C list some environment and penetration codes.

2.1.4 Conductivity and Grounding

Material conductivity plays an important role in determining the likelihood of a breakdown. The actual threat posed by internal charging depends on accumulating charge until the resulting electric field stress causes an ESD. Charge accumulation depends on retaining the charge after deposition. Since internal charging fluxes at GEO are on the order of 1 pA/cm^2 (1 pA $= 10^{-12}$ A), resistivities on the order of $10^{12}\ \Omega \cdot$ cm will conduct charge away, if grounded, so that high local electric field stress (10^5 to 10^6 V/cm) conditions cannot occur and initiate an arc. Unfortunately, modern spacecraft dielectric materials such as Teflon® and Kapton®, flame retardant 4 (FR4) circuit boards, and conformal coatings often have high enough resistivities to cause problems (Section 6.1). If the internal charge-deposition rate exceeds the leakage rate, these excellent dielectrics can accumulate charge to the point that discharges to nearby conductors are possible. If that conductor leads to or is close to a sensitive victim, there could be disruption or damage to the victim circuitry.

Metals, although conductive, may be a problem if they are electrically isolated by more than $10^{12}\ \Omega$. Some examples of metals that may be isolated (undesirable) are radiation spot shields, structures that are deliberately insulated, capacitor cans, integrated-circuit (IC) and hybrid cans, transformer cores, relay coil cans, and wires that may be isolated by design or by switches. Each of these isolated items could be an internal charging threat and should be scrutinized for its contribution to the internal charging hazards.

2.1.5 Breakdown Voltage

The *breakdown voltage* is that voltage at which the dielectric field strength of a particular sample (or air gap) cannot sustain the voltage stress and a breakdown (arc) is likely to occur. The breakdown voltage depends on the basic dielectric strength of the material [volts per mil (V/mil) is one measure of the dielectric strength] and on the thickness of the material. Even though the dielectric strength is implicitly linear, the thicker materials usually are reported to have less strength per unit thickness. Manufacturing blemishes or handling damage can all contribute to the variations in breakdown strength that will be observed in practice. As a rule of thumb, if the exact breakdown strength is not known, most common good-quality spacecraft dielectrics may break down when their internal electric fields exceed 2×10^5 V/cm (2×10^7 V/m; 508 V/mil). As a practical matter, because of sharp corners, interfaces, and vias that are inevitably present in printed circuit (PC) boards, the breakdown voltage may be less.

2.1.6 Dielectric Constant

The *dielectric constant* of a material, or its permittivity, is a measure of the electric field inside the material compared to the electric field in a vacuum. It is commonly used in the description of dielectric materials. The dielectric constant of a material (ε) is generally factored into the product of the permittivity of free space ($\varepsilon_0 = 8.85 \times 10^{-12}$ F/m) and the relative permittivity (ε_r, a dimensionless quantity) of the material in question ($\varepsilon = \varepsilon_0 \times \varepsilon_r$). Relative dielectric constants of insulating materials used in spacecraft construction generally range from 2.1 to as much as 7: assuming a relative dielectric constant of 2.7 (between Teflon® and Kapton®) is an adequate approximation if the exact dielectric constant is not known. Examples of the use of the dielectric constant for calculating time constants are given in Section E.7.

2.1.7 Shielding Density

The density of a material is important in determining its shielding properties. The penetration depth of an electron of given energy, and therefore its ability to contribute to internal charging, depends on the thickness and density of the material through which it passes. Since aluminum is a typical material for spacecraft outer surfaces, the penetration depth is commonly based on the aluminum equivalent. To the first order, the penetration depth in materials depends on the shielding mass. That is, if a material is one-half the density of aluminum, it takes twice the thickness to achieve the same shielding as that of aluminum.

2.1.8 Electron Fluxes (Fluences) at Breakdown

For IESD, the electron flux for a given duration at a location is a critical quantity. Figure 2-5 compares spacecraft disruptions as functions of environmental flux at the victim location. Experience and observations from the Combined Release and Radiation Effects Satellite (CRRES) and other satellites have shown that if the normally incident internal flux is less than 0.1 pA/cm^2, there have been few, if any, internal charging problems [2×10^{10} electrons/cm^2 in 10 h appears to be the threshold]. Bodeau [1,2] and others report problems with sensitive circuits at even lower levels on some newer spacecraft. For geosynchronous orbits, the flux above 3 MeV is usually less than 0.1 pA/cm^2, and a generally suitable level of protection can be provided by 110 mils of aluminum equivalent (Fig. 2-3).

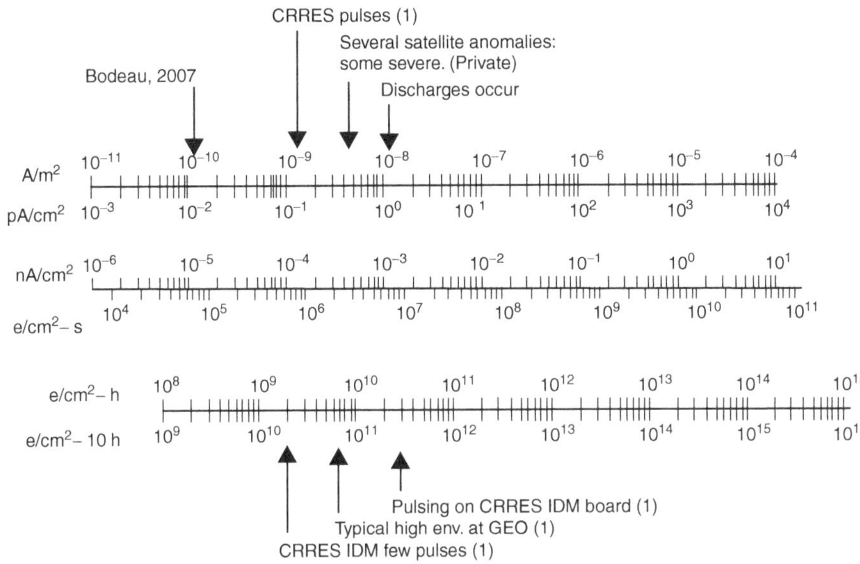

Fig. 2-5. IESD hazard levels versus electron flux (various units). The parenthetical (1)'s refer to Frederickson [3].

Modern spacecraft are being built with thinner walls or only thermal blankets (less mass), so the simple solution to the internal charging problem (adding shielding everywhere) cannot be implemented. However, adding spot shielding mass (grounded) near sensitive regions can help in many cases.

Figure 2-5 also allows a direct comparison between common units as used in the literature (i.e., 10^6 e/cm^2 · s is about 0.2 pA/cm^2). Additional information about CRRES is provided in Section B.1.2.5.

The approximation of 0.1 pA/cm^2 noted as a nominal threshold for internal charging difficulties is experientially based, not physics based, and thus has limits. Some considerations include that this is based on CRRES data (although verified by other researchers) for "typically used materials" and probably at or near room temperature. If highly resistive materials are used in cold situations and near electronics, further testing or analysis should be done.

2.2 ELECTRON ENVIRONMENT

To assess the magnitude of the IESD concern for a given orbit, it is necessary to know the electron charging environment along that orbit. (As noted earlier, the protons generally do not have enough penetrating flux to cause a significant internal charge.) The electron orbital environments of primary interest (in terms of number of affected satellites) are GEO, medium Earth orbits (MEOs), and polar Earth orbits (PEOs). Other orbital regimes that are also known to be of interest are Molniya orbits and orbits at Jupiter and Saturn (Sections B.3 and B.4).

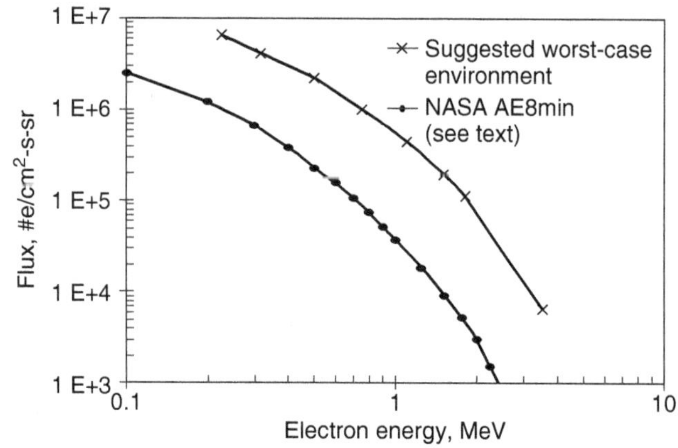

Fig. 2-6. Suggested worst-case geostationary integral electron flux environment. *Upper:* Worst-case short-term GEO environment (May 11, 1992, 197°E peak daily environment over a several hour period, with no added margin). *Lower:* NASA AE8min long-term average environment (200°E). Integral flux is for total flux greater than the energy specified.

The 11-year variation between the most severe electron environments and the least severe can vary over a 100 : 1 range and shows a correlation with the solar cycle (Section B.2.2.1, Figs. B-3 and B-4). A project manager might consider "tuning" the protection to the anticipated service period, but even in quiet years, the worst flux will sometimes be as high as the worst flux of noisy years. The environment presented in this document represents a worst-case level for GEO for any phase of the solar cycle.

Figure 2-6 shows a worst-case GEO internal charging spectrum generated by selecting dates when the Geosynchronous Operational Environmental Satellite (GOES) $E > 2$ MeV electron data values were elevated to extremely high levels and then using worst-case electron spectrum data from the geosynchronous synchronous orbit particle analyzer (SOPA) instrument for the same days. It is approximately a 99.9th percentile event (1 day in 3 years). (Sections B.1.2.3 and B.1.2.4 contain descriptions of the GOES satellite and SOPA instrument.) The GEO integral electron spectrum varies with time in both shape and amplitude. Figure 2-6 also plots the corresponding long-term nominal electron spectrum as estimated by the NASA AE8min code [4] for the same energy range. The large difference between the nominal time-averaged (AE8) and shorter-term worst-case conditions is characteristic of the radiation environment at Earth. While higher environments are less frequent, they do occur. The GEO environment varies with longitude, with a maximum flux at 200° East (Fig. B-6).

2.2.1 Units

The primary units that describe the electron environment are flux and fluence. In this book, *flux* corresponds to the rate at which electrons pass through or into a surface element. Although the units of omnidirectional flux (J) are often in terms of electrons per square centimeter ($J = 4\pi \times I$), units here will generally be the number of electrons per square centimeter per steradian (I). The time unit (per day or per second, for example) should be explicitly present. Some reports present *fluence* (flux integrated over time) but additionally describe the accumulation period (a day or 10 h, for example), which can then be converted to a flux. Electron fluxes may also be expressed as amperes (A) or picoamperes (pA) per unit area (often per cm^2). Figure 2-5 interrelates various flux and fluence units. The flux can be described as an integral over energy (electrons with energy exceeding a specified value as shown in Fig. 2-6) or differential (flux in a range of energy). ESD damage potential is related to the stored energy, which is related to fluence (flux integrated over time).

2.2.2 Substorm Environment Specifications

Worst-case plasma environments should be used in predicting spacecraft surface potentials on spacecraft. The ambient space plasma and the photoelectrons generated by solar extreme ultraviolet (EUV) are the major sources of spacecraft surface charging currents in the natural environment. The ambient space plasma consists of electrons, protons, and other ions, whose energies are described by the temperature of the plasma, as discussed in Section 2.1.1. The net current to a surface is the sum of currents caused by ambient electrons and ions, secondary electrons, photoelectrons, and other sources (e.g., ion engines, plasma contactors, and the spacecraft velocity relative to the plasma in LEO, where ram and wake effects may be present). A spacecraft in this environment accumulates surface charges until current equilibrium is reached, at which time the net current is zero. The EUV-created photoelectron emissions usually dominate in geosynchronous orbits and prevent the spacecraft potential from being very negative during sunlit portions of a mission.

The density of the plasma also affects spacecraft charging. A tenuous plasma of less than 1 particle/cm^3 will charge the spacecraft and its surfaces more slowly than will a dense plasma of thousands of particles/cm^3. Also, a tenuous plasma's current can leak off partially insulated surfaces more quickly.

Although the photoelectron current associated with solar EUV dominates over most of the magnetosphere in and near geosynchronous orbit, during geomagnetic substorms the ambient electron current can often control and dominate the charging process. Unfortunately, the ambient plasma environment at geosynchronous orbit is very difficult to describe. Detailed particle spectra (flux versus energy) are available from several missions, such as the Applications Technology Satellites (ATS-5 and ATS-6), Spacecraft Charging at High Altitudes (SCATHA), and the SOPA instruments, but these are often not easily incorporated into charging models. Rather, for simplicity, only the isotropic currents and Maxwellian temperatures are normally used by modelers; and these only for electrons and protons. Useful answers can be obtained with this simple representation. For a

Table 2-1. Worst-Case Geosynchronous Plasma Environment

Item	Units	Value	Description
N_E	cm^{-3}	1.12	Electron number density
T_E	eV	1.2×10^4	Electron temperature
N_I	cm^{-3}	0.236	Ion number density
T_I	eV	2.95×10^4	Ion temperature

worst-case static charging analysis, the single Maxwellian environmental characterization given in Table 2-1 is recommended. (Tables B-1 and B-2 and Appendix I show alternative representations of the geosynchronous orbit worst-case environments.)

Table 2-1 lists a worst-case (~90th percentile) single-Maxwellian representation of the GEO environment. Section B.1.1 describes the spacecraft charging equations and methods by which these values can be used to predict spacecraft charging effects. If the worst-case analysis shows that spacecraft surface differential potentials are less than 100 V, there should be no ESD problem. If the worst-case analysis shows a possible problem, use of more realistic plasma representations should be considered.

A more comprehensive discussion of plasma parameters is given in Section B.1.1. Alternate descriptions of plasma parameters are presented in Section B.2.1, Tables B-1 and B-2, Fig. B-1, and Appendix I, and these descriptions include fluxes and energies that might be used for material charging testing. Several original worst-case data sets for the ATS-5 and ATS-6 satellites and the SCATHA satellite, with average values, standard deviations, and worst-case values are presented in Appendix I. Additionally, percentages of yearly occurrences are given, and finally, a time history of a model substorm is provided. All of these different descriptions of plasma parameters can be used to help analyze special or extreme spacecraft charging situations. See the literature (e.g., Garrett [5], Hastings and Garrett [6], Roederer [7], and Garrett [8]) on space physics for more detailed explanations of the radiation and plasma environment.

2.3 MODELING SPACECRAFT CHARGING

Analytical modeling techniques should be used to predict surface charging effects. In this section, approaches to predicting spacecraft surface voltages resulting from encounters with plasma environments (Section 2.3.1) or high-energy particle events (Section 2.3.2) are discussed to set the stage for the charging analysis process described in subsequent sections. The descriptions are intended to provide an overview only, with the details left specifically to the appendices. Even the simple methods described, however, can be used to identify possible discharge conditions (Section 2.4) and, based on coupling models (Section 2.5), to establish the spacecraft and component-level test requirements. Again, however, for the interested reader, details are intentionally left to the appendices.

2.3.1 The Physics of Surface Charging

Although the physics behind the spacecraft charging process is quite complex, the formulation at geosynchronous orbit at least can be expressed in straightforward terms. The fundamental physical process for all spacecraft charging is that of current balance: at equilibrium (typically achieved in milliseconds for the overall spacecraft, seconds to minutes on isolated surfaces relative to vehicle ground, and up to hours between surfaces), all currents sum to zero. The potential at which equilibrium is achieved for the spacecraft is the potential difference between the spacecraft and the space plasma ground; similarly, each surface will achieve a separate equilibrium relative to space plasma and the surrounding surfaces. In terms of the ambient plasma current [9], the basic equation expressing this current balance for a uniformly conducting spacecraft at equilibrium is (see Appendix G for details)

$$\mathrm{I}_E(V) - [\mathrm{I}_I(V) + \mathrm{I}_{PH}(V) + \mathrm{I}_{\mathrm{Secondary}}(V)] = \mathrm{I}_T \tag{2.3-1}$$

where

V = spacecraft potential relative to the space plasma
I_E = incident electron current to the spacecraft surface
I_I = incident ion current to the spacecraft surface
$\mathrm{I}_{\mathrm{Secondary}}$ = additional electron currents from secondaries, backscatter, and any human-made sources; see Appendix G for details
I_{PH} = photoelectron current
I_T = total current to spacecraft (at equilibrium, $\mathrm{I}_T = 0$)

As a simple illustration of the solution of Eq. (2.3-1), assume that the spacecraft is a conducting sphere, that it is in eclipse ($\mathrm{I}_{\mathrm{PH}} = 0$), that the secondary currents are about 0, and that the plasmas are Maxwell–Boltzmann distributions. As discussed in Appendix G, the first-order currents for the electrons and ions are given by the following simple current/voltage (I/V) curves (assuming a negative potential on the spacecraft):

Electrons:

$$\mathrm{I}_E = \mathrm{I}_{E0} \exp\left(\frac{qV}{T_E}\right) \qquad V < 0 \text{ repelled} \tag{2.3-2}$$

Ions:

$$\mathrm{I}_I = \mathrm{I}_{I0}\left(1 - \frac{qV}{T_I}\right) \qquad V < 0 \text{ attracted} \tag{2.3-3}$$

where

$$\mathrm{I}_{E0} = \frac{qN_E}{2}\left(\frac{2T_E}{\pi m_E}\right)^{1/2} \tag{2.3-4}$$

$$\mathrm{I}_{I0} = \frac{qN_I}{2}\left(\frac{2T_I}{\pi m_I}\right)^{1/2} \tag{2.3-5}$$

and

N_E = density of electrons in ambient plasma (cm^{-3})

N_I = density of ions in ambient plasma (cm^{-3})

m_E = mass of electrons (9.109×10^{-28} g)

m_I = mass of ions (proton: 1.673×10^{-24} g)

q = magnitude of the electronic charge (1.602×10^{-19} C)

T_E = plasma electron temperature (eV)

T_I = plasma ion temperature (eV)

To solve the equation and find the equilibrium potential of the spacecraft relative to the space plasma, V is varied until $I_T = 0$. As a crude example, for a geosynchronous orbit during a geomagnetic storm, the potential is usually on the order of 5 to 10 kV, whereas T_I is typically about 20 to 30 keV, implying that $|qV/T_I| < 1$, so $I_I \sim I_{I0}$. Ignoring secondary currents, these approximations lead to a simple proportionality between the spacecraft potential and the ambient currents and temperatures:

$$V \sim \frac{-T_E}{q} \times Ln\left(\frac{I_E}{I_I}\right) \tag{2.3-6}$$

That is, to first order in eclipse (see, however, Appendix G), the spacecraft potential is roughly proportional to the plasma temperature expressed in electron volts and the natural log of the ratio of the electron and ion currents—a simple but useful result for estimating the order of the potential on a spacecraft at geosynchronous orbit.

To summarize, surface charge modeling is a process of computing the current balance for (1) the overall vehicle; (2) next, isolated surfaces relative to spacecraft ground; and (3) ultimately, the current flow between surfaces. An I/V relationship is determined for each surface configuration and the adjacent surfaces, and then, given the plasma environment, the potential(s) at which the current balance is achieved are computed. Clearly, this can become a complicated time-dependent process, as each electrically isolated surface on a spacecraft approaches a unique equilibrium leading to differential charging (the cause of most surface charging–generated spacecraft anomalies). Fortunately, computer codes such as Nascap-2k (Section C.3.3) have been developed that can handle very complex spacecraft configurations. See also Section C.3.4 for a description of the *Space Environments and Effects (SEE) Interactive Spacecraft Charging Handbook* tool, which is particularly useful for quickly estimating surface potentials for simple designs.

2.3.2 The Physics of Dielectric Charging

The computations involved in estimating dielectric charging resemble surface charging calculations with the inclusion of space charge. That is, the basic problem is the calculation of the electric field and charge density in a self-consistent fashion over the three-dimensional (3-D) space of interest: typically, a dielectric volume. Poisson's equation must be solved subject to the continuity equation and Ohm's law. As detailed in Section D.1, for a simple one-dimensional planar approximation, these equations (for electrons) can be reduced to a single equation where the charge buildup in a dielectric at a position x in the dielectric at time t can be described by

$$\frac{\partial(\varepsilon(x)E(x,t))}{\partial t} + \sigma(x,t)E(x,t) = -J_R(x,t) \tag{2.3-7}$$

where

E = electric field (V/cm) at x for time t
σ = conductivity $(\Omega \cdot \text{cm})^{-1} = \sigma_0 + \sigma_r$
σ_0 = dark conductivity $(\Omega \cdot \text{cm})^{-1}$
σ_r = radiation-induced conductivity $(\Omega \cdot \text{cm})^{-1}$
ε = $\varepsilon_0\varepsilon_r$ (material permittivity, F/m)
ε_0 = free-space permittivity $= 8.8542 \times 10^{-12}$ F/m
ε_r = relative dielectric constant (dimensionless)
J_R = incident particle flux (current density) where $-\partial J_R/\partial x$ = charge deposition rate at x

Note in particular that the total current consists of the incident current J_R (primary and secondary particles) and a conduction current σE driven by the electric field developed in the dielectric (the ohmic term). Integrating Eq. (2.3-7) in x across the dielectric layer then gives the variation of electric field in the dielectric at a given time. The results are stepped forward in time and the process repeated to give the changing electric field and charge density in the dielectric. As in the case of surface charging, computer codes such as NUMIT (Section C.2.7) and DICTAT (Section C.2.9) have been developed to carry out these computations and predict the buildup of electric field in the dielectric. When that field E exceeds the breakdown potential of the material, an arc discharge is possible.

2.4 DISCHARGE CHARACTERISTICS

Charged spacecraft surfaces, environmentally caused or deliberately biased, can discharge, and the resulting transients can couple into electrical systems. A spacecraft in space may be considered to be a capacitor relative to the space plasma potential. The spacecraft, in turn, is divided into numerous other capacitors by the dielectric surfaces used for thermal control and for power generation. This system of capacitors can be charged at different rates depending on incident fluxes, time constants, and spacecraft configuration effects.

The system of capacitors floats electrically with respect to the space plasma potential. This can give rise to unstable conditions in which charge can be lost from the spacecraft to space. Although the exact conditions required for such breakdowns are not known, what is known is that breakdowns do occur, and it is hoped that conditions that lead to breakdowns can be bounded. Breakdowns, or discharges, occur because charge builds up in spacecraft dielectric surfaces or between various surfaces on the spacecraft. Whenever this charge buildup generates an electric field that exceeds a breakdown threshold, charge may be released from the spacecraft to space or to an adjacent surface with a different potential. This charge release will continue until the electric field can no longer sustain an arc. Hence, the amount of charge released will be limited to the total charge stored in or on the dielectric at the discharge site. Charge loss or current to space or another surface causes the dielectric surface voltage (at least locally) to relax toward zero. Since the dielectric is coupled capacitively to the structure, the charge loss will also cause the structure potential to become less negative. In fact, the structure could become positive with respect to the space plasma potential. The exposed conductive surfaces of the spacecraft will then collect electrons from the environment (or attract back the electrons emitted) to reestablish the structure potential required by the ambient conditions. The entire process for a conducting body to charge relative to space can take only a few milliseconds; in contrast, differential charging between surfaces may take from a few seconds to hours to reach equilibrium. Multiple discharges can be produced if intensities remain high long enough to reestablish the conditions necessary for discharge.

It is well known in the spacecraft solar array community that there can be a charge loss over an extended area of the dielectric (NASA TP-2361) [10]. This phenomenon is produced by the plasma cloud from a discharge sweeping over dielectric surfaces where the underlying conductor is connected electrically to the arc site. Charge loss from solar array arcs has been seen for distances of 2 m and more from the arc site and can involve capacitances of several hundreds of picofarads in the discharge, depending on configuration. This phenomenon can produce area-dependent charge losses capable of generating currents of 4 to 5 A. The differential voltages necessary to produce this large charge-clean-off type of discharge may be as low as 1000 V on solar arrays, depending on the specific type of array, geometric configuration, or environment. In modeling the charged surfaces swept free of charge by an arc, one should assume that all areas with substrates connected electrically directly to the arc site and with a line of sight to the arc site will be discharged, and calculate the arc energy accordingly.

Because sunlight tends to charge all illuminated surfaces a few volts positive relative to the ambient plasma and shaded dielectric surfaces may charge strongly negatively, differential charging is likely to occur between sunlit and shadowed surfaces. Since breakdowns are believed to be related to differential charging, they can occur during sunlit charging events. Entering and exiting an eclipse, in contrast, results in a change in absolute charging for all surfaces except those weakly capacitively coupled to the structure (capacitance to structure less than that of spacecraft to space, normally $< 2 \times 10^{-10}$ F). Differential charging in eclipse develops slowly and depends on differences in secondary yield. In the following paragraphs, each of the breakdown mechanisms identified is summarized.

2.4.1 Dielectric Surface Breakdowns

If either of the following criteria is exceeded, discharges can occur:

1. If electric fields reach a magnitude that exceeds the breakdown strength of the surrounding "empty" space, a discharge may occur [11]. A published rule of thumb [12] is that if dielectric surface voltages resulting from spacecraft surface charging are greater than about 500 V, positive relative to an adjacent exposed conductor, a breakdown may occur. In this document we have adopted a more conservative 400 V differential voltage threshold of concern for ESD breakdown. This is not true for induced potentials such as from solar arrays or Langmuir probes; these should be analyzed separately. The physics of electric field breakdown in gases has been explained by Townsend (see, e.g., reference [11]).
2. The interface between a visible surface dielectric and an exposed grounded conductor has an electric field greater than 10^5 V/cm (NASA TP-2361) [10]. Note that edges, points, gaps, seams, and imperfections in surface materials can increase electric fields and hence promote the probability of discharges. These items are not usually modeled and must be found by close inspection of the exterior surface specifications. In some cases, a plasma cloud generated by micrometeorid/debris impact at the site could trigger the breakdown.

The first criterion can be exceeded by solar arrays in which the high secondary yield of the cover slide can result in surface voltages that are positive with respect to the metallized interconnects. This criterion can also apply to metallized dielectrics in which the metallized film, either by accident or design, is isolated from structure ground by a large or nonexistent resistance (essentially only capacitively coupled). In the latter case, the dielectric can be charged to large negative voltages (when shaded), and the metal film will thus become more negative than the surrounding surfaces and act as a cathode or electron emitter.

In both of these conditions, stored charge is initially ejected to space in the discharge process. This loss produces a transient that can couple into the spacecraft structure and possibly into the electronic systems. Current returns from space to the exposed conductive areas of the spacecraft. Transient currents flow in the structure depending on the electrical characteristics. It is assumed that the discharge process will continue until the voltage gradient or electric field that began the process disappears. The currents flowing in the structure will damp out according to its resistance.

The computation of charge lost in any discharge is highly speculative at this time. Basically, charge loss can be considered to result from the depletion of two capacitors: that stored in the spacecraft, which is charged to a specified voltage relative to space, and that stored in a limited region of the dielectric at the

Table 2-2. Rough Magnitudes of Surface ESD Event Parameters

Size	$Q(C)$	$C(F)$	V	$E(J)$
Minor, up to	500 nC	500 pF	1 kV	250 μJ
Moderate, up to	2 μC	500 pF	4 kV	4 mJ
Severe, up to	10 μC	500 pF	20 kV	100 mJ

discharge site. The prediction of charge loss requires not only the calculation of voltages on the spacecraft, but a careful review by an experienced analyst as well.

As a guide, the following charge loss categories are useful (as adapted from NASA TP-2361 [10]):

$$0 < Q_{\text{lost}} < 0.5\ \mu\text{C} \qquad \text{minor discharge}$$

$$0.5 < Q_{\text{lost}} < 2\ \mu\text{C} \qquad \text{moderate discharge}$$

$$2 < Q_{\text{lost}} < 10\ \mu\text{C} \qquad \text{severe discharge}$$

Energy, voltage, or discharge considerations can also be quantified as a means to characterize the severity of a dielectric discharge (or discharge from an isolated conductor). Assuming a 500 pF discharge capacitance as default and using the Q_{lost} criteria above in standard equations yields Table 2-2.

2.4.2 Buried (Internal) Charge Breakdowns

In this section we describe the situation where charges have sufficient energy to penetrate slightly below the surface of a dielectric and are trapped. If the dielectric surface is maintained near zero potential because of photoelectron or secondary electron emission, strong electric fields may exist in the material. This can lead to electric fields inside the material large enough to cause breakdowns. Breakdown can occur whenever the internal electric field exceeds 2×10^5 V/cm (2×10^7 V/m, ~508 V/mil). Table 6.1 lists the breakdown strength of some dielectric materials.

A layer of charge with 2.2×10^{11} e/cm^2 will create a 2×10^7 V/m electric field in a material with a relative dielectric constant of 2. (The E-field is proportional to charge and inversely proportional to the dielectric constant.)

2.4.3 Spacecraft-to-Space Breakdowns

Spacecraft-to-space breakdowns are generally similar to dielectric surface breakdowns but involve only small discharges. It is assumed that a strong electric field exists on the spacecraft surfaces—usually because of a geometric interfacing of metals and dielectrics. Periodically, this arrangement triggers a breakdown of the spacecraft capacitor. Since this spacecraft-to-space capacitance tends to be on the order of 2×10^{-10} F, these breakdown transients should be small and rapid. Based on an assumption of a 2 kV breakdown, the resulting stored energy is minor, in accordance with Table 2-2.

2.5 COUPLING MODELS

Coupling model analyses must be used to determine the hazard to electronic systems from exterior discharge transients. In this section, techniques for computing the influence of exterior discharge transients on interior spacecraft systems are discussed.

2.5.1 Lumped-Element Modeling

Lumped-element models (LEMs) have been used to define the surface charging response to environmental fluxes [13–16] and are currently used to predict interior structural currents resulting from surface discharges. The basic philosophy of a LEM is that spacecraft surfaces and structures can be treated as electrical circuit elements: resistance, inductance, and capacitance. The geometry of the spacecraft is considered only to group or lump areas into nodes within the electrical circuit in much the same way as surfaces are treated as nodes in thermal modeling. These models can therefore be made as simple or as complex as is considered necessary for the circumstances.

The LEMs for discharges assume that the structure current transient is generated by capacitive coupling to the discharge site and is transmitted in the structure by conduction only. An analog circuit network is constructed by taking into consideration the structure properties and the geometry. This network must consider the principal current flow paths from the discharge site to exposed conductor areas: the return path to space plasma ground. Discharge transients are initiated at regions in this network selected as being probable discharge sites by surface charging predictions or other means. Choosing values of resistance, capacitance, and inductance to space can control transient characteristics. Network computer transient circuit analysis programs such as ISPICE, the first commercial version of SPICE (Simulation Program with Integrated Circuit Emphasis), and SPICE2 can solve the resulting transients within the network.

LEMs developed to predict surface charging rely on the use of current input terms applied to surfaces independently. Since there are no terms relating the influence of charging by one area on the incoming flux to other areas, the predictions usually result in larger negative voltages than are actually observed. Other modeling techniques that take these three-dimensional effects into account, such as is done in Nascap-2k (Section C.3.3), predict surface voltages closer to those measured. Here, Nascap-2K is the currently recommended analysis technique for surface charging.

2.5.2 Electromagnetic Coupling Models

Numerous programs have been developed to study the effects of electromagnetic coupling on circuits. Such programs have been used to compute the effects of an electromagnetic pulse and that of an arc discharge. One program, the Specification and Electromagnetic Compatibility Program (SEMCAP) [17] developed by

TRW Inc. (now Northrop-Grumman Space Technology), has successfully analyzed the effects of arc discharges on an actual spacecraft, *Voyager*.

REFERENCES

[1] M. Bodeau, "Going Beyond Anomalies to Engineering Corrective Action, New IESD Guidelines Derived from a Root-Cause Investigation," *2005 Space Environmental Effects Working Group Workshop*, Aerospace Corporation, El Segundo, California, 2005. See also Balcewicz et al. [19].

[2] M. Bodeau, "High Energy Electron Climatology That Supports Deep Charging Risk Assessment in GEO," *48th AIAA Aerospace Sciences Meeting*, Orlando, Florida, AIAA 2010-1608, 2010. A fine work with good concepts, explained and illustrated with actual space data and estimates of fluence accumulation versus material resistivity. Bodeau challenges the 0.1-pA/cm^2 and 10-h flux integration guidelines.

[3] A. R. Frederickson, E. G. Holeman, and E. G. Mullen, "Characteristics of Spontaneous Electrical Discharges of Various Insulators in Space Radiation," *IEEE Transactions on Nuclear Science*, vol. 39, no. 6, pp. 1773–1782, December 1992. This document is a description of the best-known attempt to quantify internal charging effects on orbit by means of a well-thought-out experiment design. The results were not all that the investigators had hoped for, but the data are excellent and very good conclusions can be reached from them despite the investigators' concerns.

[4] J. I. Vette, *The AE-8 Trapped Electron Model Environment*, NSSDC/WDC-A-R&S, Report 91-24, National Space Science Data Center, Goddard, Maryland, November 1991.

[5] H. B. Garrett, "Review of Quantitative Models of the 0 to 100 keV Near-Earth Plasma," *Reviews of Geophysics and Space Physics*, vol. 17, no. 3, pp. 397–416, 1979.

[6] D. Hastings and H. B. Garrett, *Spacecraft–Environment Interactions*, Atmospheric and Space Science Series, A. J. Dessler, Ed., Cambridge University Press, Cambridge, England, 292 pages, 1996. This is a college textbook on the subject. It contains additional reading for the space environment and interactions with spacecraft in its Section 4.

[7] J. G. Roederer, *Dynamics of Geomagnetically Trapped Radiation*, Springer-Verlag, New York, 166 pages, 1970.

[8] H. B. Garrett, *Guide to Modeling Earth's Trapped Radiation Environment*, AIAA G-083-1999, American Institute of Aeronautics and Astronautics, Reston, Virginia, 55 pages, 1999.

[9] H. B. Garrett, "The Charging of Spacecraft Surfaces," *Reviews of Geophysics and Space Physics*, vol. 19, no. 4, pp. 577–616, November 1981. A nice summary paper, with numerical examples and many illustrations. This and Whipple [20] are two definitive papers on the subject, each covering slightly different aspects.

[10] C. K. Purvis, H. B. Garrett, A. C. Whittlesey, and N. J. Stevens, *Design Guidelines for Assessing and Controlling Spacecraft Charging Effects*, NASA Technical Paper 2361, National Aeronautics and Space Administration, September 1984. This document has been widely used by practitioners of this art (usually, EMC engineers or radiation survivability engineers) since its publication in 1984. Its contents are limited to surface charging effects. The contents are valid to this day for that purpose. NASA TP-2361 contents have been incorporated into this and NASA-STD-4002, Revision A, with

heavy editing. Many of the original details, especially time-variant and multiple-case versions of suggested environments, have been simplified into single worst-case environments in NASA-HDBK-4002, Revision A. Some background material has not been transferred into this document, so the original may still be of interest.

[11] M. S. Naidu and V. Kamaraju, *High Voltage Engineering, 4th ed.*, Tata and McGraw-Hill Publishing Company Limited, New Delhi, India, 2009. Townsend's criteria for breakdown are described in this book.

[12] P. Coakley, *Assessment of Internal ECEMP with Emphasis for Producing Interim Design Guidelines*, AFWL-TN-86-28, Air Force Weapons Laboratory, June 1987. This 63-page document (produced by JAYCOR, now part of L-3 Communications) was excellent in its day and still is a fine reference that should be on every ESD practitioner's bookshelf. The design numbers from that document are very similar to those of this book.

[13] P. A. Robinson, Jr. and A. B. Holman, "Pioneer Venus Spacecraft Charging Model," *Proceedings of the Spacecraft Charging Technology Conference*, AFGL-TR-77-0051/NASA TMX-73537, National Aeronautics and Space Administration, pp. 297–308, 1977.

[14] G. T. Inouye, "Spacecraft Potentials in a Substorm Environment," *Spacecraft Charging by Magnetospheric Plasma*, Vol. 42, A. Rosen, Ed., MIT Press, Cambridge, Massachusetts, pp. 103–120, 1976.

[15] M. J. Massaro, T. Green, and D. Ling, "A Charging Model for Three-Axis Stabilized Spacecraft," *Proceedings of the Spacecraft Charging Technology Conference*, AFGL-TR-77-0051/NASA TMX-73537, National Aeronautics and Space Administration, pp. 237–270, 1977.

[16] M. J. Massaro and D. Ling, "Spacecraft Charging Results for the DSCS-III Satellite," *Spacecraft Charging Technology Conference*, Air Force Geophysics Laboratory, Hanscom Air Force Base, Massachusetts, NASA Conference Publication 2071/AFGL TR-79-0082, National Aeronautics and Space Administration, pp. 158–178, 1979.

[17] R. Heidebrecht, *SEMCAP Program Description*, Version 7.4, TRW Electromagnetic Compatibility Department, Space Vehicles Division, TRW Systems Group, Redondo Beach, California, 1975.

[18] M. J. Berger, J. S. Coursey, M. A. Zucker, and J. Chang, *Stopping-Power and Range Tables for Electrons, Protons, and Helium Ions*, NISTIR 4999, National Institute of Standards and Technology; also available at http://www.nist.gov/physlab/data/star, April 26, 2010.

[19] P. Balcewicz, J. M. Bodeau, M. A. Frey, P. L. Leung, and E. J. Mikkelson, "Environmental On-Orbit Anomaly Correlation Efforts at Hughes," *6th Spacecraft Charging Technology Conference Proceedings*, AFRL-VS-TR-20001578, pp. 227–230, 1998.

[20] E. C. Whipple, "Potentials of Surfaces in Space," *Reports on Progress in Physics*, vol. 44, pp. 1197–1250, 1981. Nice summary paper. Emphasis is on total charging and not internal charging, but good for physics background. This and Garrett [9] are the two definitive papers on that subject, each covering slightly different aspects.

3 Spacecraft Design Guidelines

In Section 3.1 we describe processes involved in immunizing a spacecraft against spacecraft charging problems. Design guidelines are listed in Section 3.2. If the reader wishes to make a requirements document, the basic requirements include:

1. Determine whether or not the mission passes through or stays in regions with a charging threat.
2. If in a charging threat region, determine the threat that is applicable to that environment.
3. Implement measures to mitigate the threat to an acceptable level.

Sections 3.2.1 (General ESD Design Guidelines), 3.2.2 (Surface ESD Design Guidelines, Excluding Solar Arrays), 3.2.3 (Internal ESD Design Guidelines), 3.2.4 (Solar Array ESD Design Guidelines), and 3.2.5 (Special Situations ESD Design Guidelines) can be used as aids.

3.1 PROCESSES

The system developer should demonstrate through design practices, test, and analysis that spacecraft charging effects will not cause a failure to meet mission objectives. We discuss those processes briefly in this section.

3.1.1 Introduction

The classic approach to avoiding or eliminating electromagnetic problems is to look at the source of the problem, the victim, and the coupling between them. In the case of space charging, excess electrons deposit on surface or external spacecraft areas or penetrate directly to victim circuit areas, the charge being buried in a circuit board immediately adjacent to the victim. As a result, the three elements (source, coupling, and victim) are not always clearly distinguishable. For that reason, this book has disregarded these categories; however, this approach may sometimes be fruitful and is described below for completeness.

Guide to Mitigating Spacecraft Charging Effects, First Edition. Henry B. Garrett, Albert C. Whittlesey.

*3.1.1.1 **Source.*** The basic source of in-space charging problems is the charged particle environment (CPE). If that environment cannot be avoided, the next sources of ESD threats are items that can store and accumulate charge and/or energy. Ungrounded (isolated) metals are hazardous because they can accumulate charge and energy. Excellent dielectrics can accumulate charge and energy as well. Limiting the charge storing material or charging capacity is a useful method for reducing the internal charging threat. This can be accomplished by providing a bleed path so that all plasma-caused charges can equalize throughout the spacecraft or by having only small quantities of charge-storing materials.

*3.1.1.2 **Coupling.*** Coupling energy from a source via a spark (ESD) is very configuration dependent and a function of the radiated and directly coupled signals. An ESD can occur in a variety of ways, such as from metal to metal, metal to space, metal to dielectric, dielectric to dielectric, or through dielectric breakdown. The configuration of the charges determines the type of breakdown and hence the form of coupling. An isolated conductor can discharge directly into an IC lead, causing serious physical damage at the site, or the arc can induce an attenuated signal into a nearby wire, causing little damage but inducing a spurious signal. As these examples illustrate, the coupling must thus be estimated uniquely for each situation. Eliminating coupling paths from a spark source to a victim will significantly lower the ESD threat. Coupling paths could be eliminated by separation, shielding, or filtering.

*3.1.1.3 **Victim.*** A victim is any part, component, subsystem, or element of a spacecraft that can be affected adversely by an arc discharge (or field effects, in the case of some science instruments). Given the different effects of ESDs, the types and forms of victims can be highly variable. ESD and electromagnetic compatibility (EMC)-induced parts failures, although major problems, are not the only ones. Effects can range from the so-called soft errors (e.g., a memory element may be reset) to actual mechanical damage, where an arc actually destroys material. Thus, the victims can range from individual parts to entire systems, from electronic components to optical parts. (Discharging in glass has long been known to cause fracturing and damage to optical windows or dielectrics, but empirical data suggest that optical lenses have apparently had largely successful usage in space.) The major victims and design sources will be either individual electronic components or cables that can couple the transient voltage into a subsystem. Shielding or filtering at the victim will limit the adverse effects of ESDs.

3.1.2 Design

The designer should be aware of design guidelines to avoid surface and internal charging problems (Sections 3.2.2 and 3.2.3). All guidelines should be considered in the spacecraft design and applied appropriately to the given mission.

3.1.3 Analysis

Analysis should be used to evaluate a design for charging in the specified orbital environment. There are two major approaches to such an analysis: a simple analysis and a detailed analysis, perhaps with a computer code. A very simple analysis of internal electrostatic charging is illustrated in Appendix D. Several appropriate computer transport codes are listed in Section C.2. An example of a simple surface charging analysis is described in Appendix G.

3.1.4 Testing and Measurement

Testing usually ranks high among the choices to verify and validate the survivability of spacecraft hardware in a given environment. For spacecraft charging environments, it is difficult to replicate the actual energetic plasma and total threat in all respects. The real electron environment can envelop the entire spacecraft and has a spectrum of energies. There is no test facility that can replicate all the features of that environment. As a consequence, verification and validation of charging protection are done with lower-level hardware tests and with less realistic test environments. This does not reduce the value of the tests, but additional analyses must be done to provide design validation where testing alone is inadequate. Several tests that can be performed to validate different aspects of charging are described briefly below.

3.1.4.1 Material Testing. Material electrical properties should be known before they are used. The key material properties needed are the ability to accumulate charge (i.e., resistivity or conductivity) and the pulse threat (e.g., stored voltage: energy). Secondary but important parameters include resistivity changes with time in space, temperature (cold is more resistive), and to a lesser extent, radiation- and E-field-induced conductivity. Other properties are secondary electron emission, backscatter emission, and photoelectron emission properties. Surface contamination of materials in space also changes their charging behavior.

Information about these parameters can be obtained from reference texts or by electron beam tests or conventional electrical tests. (Section 6.1 contains a sample dielectric materials list.) Analysis or tests can be used to determine the threat for particular sizes and shapes of these materials. Some test methods are described in Appendix E.

3.1.4.2 Circuit/Component Testing. The susceptibility threshold of components (i.e., transistors, ICs, etc.) is useful in understanding the threat from ESD events. The susceptibility can be a disruption threshold or a damage threshold. A Vzap test (Section E.8) can be used to determine an electronic device's capability to withstand the effects of an electrical transient.

3.1.4.3 Assembly Testing. Potentially susceptible assemblies should be tested for sensitivity to ESD. The assembly to be tested is to be mounted on a baseplate and tested while operating. Pulses are to be injected through the box of the assembly or injected into the pins of the connector while the performance of the assembly is monitored for upsets. The pulses used are to cover the expected range of current amplitudes, voltages, and pulse durations. It is very important that the pulse injection device be isolated electrically from the assembly being tested and the monitoring equipment. It is also important to ensure that the transient is not disturbing the support equipment.

3.1.4.4 System Testing. System-level testing is often the final proof that a system can survive a given environment. For IESD environments, system testing is not feasible. Materials, circuit, and assembly testing, together with analysis, must provide the system-level verification for internal charging concerns.

3.1.5 Inspection

Inspection is an important means for recognizing and minimizing the possibility of spacecraft charging discharge-induced anomalies. This inspection should be conducted as the spacecraft is being assembled by a person experienced in recognizing likely areas of concern from environmentally induced interactions. A list of acceptable values of resistance for joints and connections within the spacecraft should be generated ahead of the inspection, but the inspector should take a broader view and look for other possible areas of concern.

3.2 DESIGN GUIDELINES

This section contains general guidelines and quantitative recommendations on design guidelines/techniques that should be followed in hardening spacecraft systems to spacecraft charging effects. This section contains design guidelines divided into subsections: General (Section 3.2.1), Surface Charging (Section 3.2.2), Internal Charging (Section 3.2.3), Solar Arrays (Section 3.2.4), and Special Situations (Section 3.2.5).

3.2.1 General ESD Design Guidelines

3.2.1.1 Orbit Avoidance. If possible, avoid orbits and altitudes where charging is an issue. Usually, this is not an option (Figs. 1-1 and 1-2 show hazardous environments near Earth).

3.2.1.2 Shielding. Shield all electronic elements with sufficient aluminum-equivalent thickness so that the internal charging rate is benign. Experience has shown that for GEO orbits and today's hardware, an adequate shielding level has been on the order of 110 mils of aluminum-equivalent shielding, but 200 mils is more conservative and may be necessary for certain situations

(Section 3.2.3.2.2). For some ESD-immune hardware, the amount needed may be less; it almost certainly will exceed 33 mils but may be as low as 70 mils. This is the total shielding, accounting for geometry. A more accurate determination can be done by ray tracing using radiation shielding codes capable of handling detailed geometric and spacecraft material descriptions, and comparing results to the sensitivity of possible victims.

Shield all electronic elements in a Faraday cage construction. The primary spacecraft structure, electronic component enclosures, and electrical cable shields should provide a physically and electrically continuous shielded surface around all electronics and wiring (Faraday cage). The primary spacecraft structure should be designed as an electromagnetic-interference (EMI)-tight shielding enclosure (Faraday cage). The purposes of the shielding are to prevent entry of charged particles into the spacecraft interior and to shield the interior electronics from the radiated and conducted noise of an electrical discharge on the exterior of the spacecraft. All shielding should provide at least 40-dB attenuation of radiated electromagnetic fields associated with surface discharges. An approximate 40 mil thickness of aluminum or magnesium will easily provide the desired attenuation if made electromagnetically tight. This enclosure should be as free of holes and penetrations as possible. Many penetrations can be closed by use of well-grounded metallic meshes and plates. All openings, apertures, and slits should be eliminated to maintain the integrity of the Faraday cage.

The vacuum-deposited aluminum metallization on multilayer insulation thermal blankets is insufficient to provide adequate shielding for both EMC and internal charging. Layers of aluminum foil mounted to the interior surface and properly grounded can be used to increase the shielding effectiveness of blankets or films. Aluminum honeycomb structures and aluminum face sheets can also provide significant attenuation. Electronic enclosures and electrical cables exterior to the main Faraday cage region should also be shielded to extend the coverage of the shielded region to 100% of the electronics. Unless all seams, penetrations, gaps, and so on, are shielded with a totally connected conductive skin, the Faraday cage implementation is incomplete and cannot be counted as proper protection to the interior electronics. For example, a viewing aperture of a star tracker is a penetration. Another example is a "mouse hole" for cable penetrations. All must be given careful attention as to the effects of the violation of the Faraday cage principle.

Cable shields exterior to the Faraday cage are used to maintain and extend the cage region from their exit from or entrance to the main body of the spacecraft. Cable shields should be fabricated from aluminum or copper foil, sheet, or tape. Standard coaxial shielding or metallized plastic tape wraps on wires do not provide adequate shielding protection for internal charging protection and should not be used. Shields should be terminated when they enter the spacecraft structure from the outside and carefully grounded at the entry point with a 360° EMC connector. Braid shields on wires should be soldered to any overall shield wrap and grounded at the entrances to the spacecraft. Conventional shield grounding through a connector pin to a spacecraft interior location cannot be used without violating the total shielding integrity.

Electrical terminators, connectors, feedthroughs, and externally mounted components (such as diodes) should be electrically shielded, and all shielded connector covers must be bonded to the common structural ground of the space vehicle.

3.2.1.3 Bonding. Bond all structural elements. Identify isolated conducting elements and provide bonding to chassis for those areas. Make a separate bond strap for conductive elements mounted at the end of dielectric booms. Every conductive internal part should be connected by a deliberate or leakage impedance to chassis as measured with an ohmmeter; 10^{12} Ω in a vacuum is adequate. A design with a leakage resistance below 10^{8} Ω permits construction verification with a good handheld ohmmeter. Conductive fittings on dielectric structural parts should also comply.

All conducting elements, surface and interior, should be related to a common electrical ground reference, either directly, through a charge bleed-off resistor, or via a controlled voltage on the conductor as in electrical/electronic circuitry (nothing electrically floating). All structural and mechanical parts, electronics boxes, enclosures, and so on, of the spacecraft should be electrically bonded to each other. All principal structural elements should be bonded by methods that assure a direct-current (dc) resistance of less than 2.5 mΩ at each joint if required for EMC or electrical ground referencing reasons; otherwise, a high-value bleed resistance is permissible. The collection of electrically bonded structural elements is referred to as *structure* or *structure ground*. The objective is to provide a low-impedance path for any ESD-caused currents that may occur and to provide an excellent ground for all other parts of the spacecraft needing grounding. If structure ground reference must be carried across an articulating joint or hinge, a ground strap, as short as possible, should carry the ground across the joint. Relying on bearings for a ground path is unacceptable. If structural ground must be carried across slip rings on a rotating joint, at least two (preferably more) slip rings should be dedicated to the structural ground path, some at each end of the slip ring set. The bond to structure should be achieved within 15 cm of the slip ring on each end of the rotating joint. Slip rings chosen for grounding should be remote from any slip rings carrying sensitive signals.

3.2.1.3.1 Surface Materials and Their Bonding. All spacecraft surface (visible, exterior) materials should be conductive in an ESD sense (Section 3.2.1.5). All such conductive surface materials should be electrically bonded (grounded) to the spacecraft structure. Because they are intended to drain space-charging currents only, the bonding requirements are less stringent than those for structural bonding. The dc impedance to structure should be compatible with the surface resistivity requirements; that is, less than about 10^{9} Ω from a surface to structure. The dc impedance must remain less than 10^{9} Ω over the service life of the bond in vacuum, under temperature, under mechanical stress, and so on.

3.2.1.3.2 Wiring and Cable Shields and Their Bonding. All wiring and cabling entering or exiting the shielded Faraday cage portion of the spacecraft (Section 3.2.1.2) must be shielded. Those cable shields and any other cable shields used for ESD purposes must be bonded (grounded) to the Faraday cage at the entry to the shielded region as follows:

1. The shield must be terminated 360° around a metal-shielded backshell, which in turn must be terminated to the chassis 360° around the cabling.
2. The shield bond (ground) should not be terminated by using a connector pin that penetrates the Faraday cage and receives its ground inside the shielded region.
3. A mechanism should be devised that automatically bonds the shield to the enclosure/structure ground at the connector location, or a ground lug that uses less than 15 cm of ground wire should be provided for the shield, and procedures which verify that the shield is grounded at each connector mating should be established.
4. The other end of the cable shield should be terminated in the same manner. The goal is to maintain shielding integrity even when some electronics units must be located outside the basic shielded region of the spacecraft.

3.2.1.3.3 Electrical and Electronic Grounds. Signal and power grounds (zero-volt reference points) require special attention in the way they are connected to the spacecraft structure ground. NASA-HDBK-4001, *Electrical Grounding Architecture for Unmanned Spacecraft* [1], is a good reference. For ESD purposes, a direct wiring of electrical/electronics units to structure is most desirable. In particular, do not use separate ground wires daisy-chained from unit to unit or from each unit to a distant single point (star ground) on the structure.

3.2.1.4 Conductive Path. Have a conductive path to the structure for all circuitry. A simple and direct ground path is preferred without outside wiring to the ground point. Note areas where circuits or wires may be isolated for any reason. Place bleed resistors on all circuit elements that may become unreferenced (floating) during mission events, such as switching or connector demating. Use NASA-HDBK-4001 [1] as a guide to eliminate ground loops if necessary.

3.2.1.5 Material Selection. Limit usage of excellent dielectrics. Metals are conductive, and protecting them from internal charging is a relatively simple matter of ensuring a charge-leakage path. Therefore, the materials of concern in controlling internal charging are dielectrics. Prominent dielectrics in modern satellites include, but are not limited to, Teflon®, Kapton®, and FR4 circuit boards. These are excellent charge-storing materials. Their use should be avoided if possible, especially in large blocks. Usages such as wire insulation or thin films (5 mils, for example) seem to contribute fewer or no problems on the interior of spacecraft. Circuit board materials may be a problem, but densely populated boards are less of a problem; short paths through the dielectric to nearby circuit traces permit easy electron bleed-off. Validate material performance with electron-beam tests in accordance with Section E.1. Brunson and Dennison [2] have measured dielectric resistivity at lower temperatures and quantified the known increase in resistivity with decreasing temperature.

Make all interior dielectrics electrically leaky. Internal dielectrics should be static-dissipative or leaky. This applies specifically to circuit boards but would be desirable for all dielectrics, including cable wiring and conformal coatings. The degree of leakiness or conductivity does not need to be great enough to interfere with circuit performance. It can be on the order of 10^4 to 10^{11} $\Omega \cdot$cm or of 10^5 to 10^{12} Ω/square (see Section E.3 for a discussion of Ω/square) and still provide a bleed path to electrons for internal charging purposes.Verify that the conductivity remains adequate over the mission life. Meeting this requirement and also providing the other necessary properties (i.e., mechanical, workable, etc.) may be a challenge.

Make all spacecraft exterior surfaces at least partially conductive. The best way to avoid differential charging of spacecraft surfaces is to make all surfaces conductive and bonded to the spacecraft structure. However, typical spacecraft surface materials often include insulating materials such as Mylar, Kapton, Teflon, fiberglass, glass, quartz, or other excellent dielectric. It should be recognized in the design phase that there may be areas for which the use of dielectric surfaces is particularly crucial, such as areas adjacent to receivers or antennas operating at less than 1 GHz, sensitive detectors (Sun and Earth detectors, etc.), or areas where material contamination or thermal control is critical. For these applications, use of (grounded) indium tin oxide (ITO) coatings is recommended.

In this section we first define the conductivity requirements for spacecraft surface materials. Materials that are typically used are then evaluated, and their usage is discussed. Analysis is suggested to estimate the effects of any dielectric surfaces that may remain on the spacecraft. At the conclusion of this section, use of materials with a high secondary electron yield is discussed.

3.2.1.5.1 Surface Material Selection Advice. By the proper choice of available materials, the differential charging of spacecraft surfaces can be minimized. At present, the only proven way to eliminate spacecraft potential variations is by making all surfaces conductive and connecting them to a common ground. Surface coatings in use for this purpose include conductive conversion coatings on metals, conductive paints, and transparent, partially metallic vacuum-deposited films, such as ITO. Table 3-1 describes some of the more common acceptable surface coatings and materials with a successful use history. Table 3-2 describes other common surface coatings and materials that should be avoided if possible.

Table 3-1. Surface Coatings and Materials Acceptable for Spacecraft Use[a]

Material	Comments
Paint (carbon black)	Work with manufacturer to obtain paint that satisfies ESD conductivity requirements of Section 3.2.2 and thermal, adhesion, radiation tolerance, and other needs.
GSFC NS43 paint (yellow)	Has been used in some applications where surface potentials are not a problem; apparently will not discharge.
ITO (250 nm)	Can be used where some degree of transparency is needed; must be grounded properly. For use on solar cells, optical solar reflectors, and Kapton film, use sputtered method of application and not vapor deposited.
Zinc orthotitanate paint (white ZOT)	Possibly the most conductive white paint; adhesion difficult without careful attention to application procedures, and then difficult to remove.
Alodyne	Conductive conversion coatings for magnesium, aluminum, etc., are acceptable.
DuPont Kapton XC family	Carbon-filled polyimide films; 100XC10E7 with nominal resistivity of $2.5 \times 10^4\ \Omega \cdot \text{cm}$; not good in atomic oxygen environment without protective layer (ITO, for example).
Deposited conductors	Examples: aluminum, gold, silver, Inconel on Kapton, Teflon, Mylar, and fused silica.
Conductive paints	Over dielectric surfaces, with some means to assure bleed-off of charge.
Carbon-filled Teflon or Kapton	Carbon filler helps make the material conductive.
Conductive adhesives	Especially if needed for bridging between a conductor and ground.
Conductive surface materials	Graphite epoxy (scuffed to expose carbon fibers) or metal.
Etched metal grids	Etched or bonded to dielectric surfaces, frequent enough to have surface appear to be grounded.
Aluminum foil or metallized plastic film tapes	If they can be tolerated for other reasons, such as thermal behavior.

[a]Must be grounded to the chassis.

The following materials have been used to provide conducting surfaces on spacecraft (remember, these conductive surfaces must be grounded or at least not floating):

1. Vacuum-metallized dielectric materials in the form of sheets, strips, or tiles. The metal-on-substrate combinations include aluminum, gold, silver, and Inconel on Kapton, Teflon, Mylar, and fused silica.

Table 3-2. Surface Coatings and Materials to Be Avoided for Spacecraft Use

Material	Comments
Anodyze	Anodyzing produces a high-resistivity surface to be avoided for ESD applications. The coating can be made quite thin and might be acceptable if analysis shows that the stored energy is small.
Fiberglass material	Resistivity is too high and is worse at low temperatures.
Paint (white)	In general, unless a white paint is measured to be acceptable, it is unacceptable.
Mylar (uncoated)	Resistivity is too high.
Teflon (uncoated)	Resistivity is too high. Teflon has demonstrated long-time charge storage ability and causes catastrophic discharges.
Kapton (uncoated)	Generally unacceptable because of high resistivity; however, in continuous sunlight applications if less than 0.13 mm (5 mils) thick, Kapton is sufficiently photoconductive for use.
Silica cloth	Has been used for antenna radomes. It is a dielectric, but because of numerous fibers or if used with embedded conductive materials, ESD sparks may be individually small. It has particulate issues, however.
Quartz and glass surfaces	It is recognized that solar cell cover slides and second-surface mirrors have no substitutes that are ESD acceptable; they can be ITO-coated with minor performance degradation, and the ITO must be grounded to chassis. Their use must be analyzed and ESD tests performed to determine their effect on neighboring electronics. Be aware that low temperatures significantly increase the resistivity of glasses [3].

2. Thin, conductive front-surface coatings, especially ITO on fused silica, Kapton, Teflon, or dielectric stacks.
3. Conductive paints, fog (thin paint coating), carbon-filled Teflon, or carbon-filled polyester on Kapton (Sheldahl black Kapton).
4. Conductive adhesives.
5. Exposed conductive facesheet materials (graphite/epoxy scuffed with fine sandpaper to expose conductive graphite fibers, or metal).
6. Etched metal grids or bonded (or heat embedded) metal meshes on non-conductive substrates.
7. Aluminum foil or metallized plastic film tapes.

Because of the variety in the configuration and properties of these materials, there is a corresponding variety in the applicable grounding techniques and specific concerns that must be addressed to ensure reliable in-flight performance.

The following practices have been found useful for grounding/bonding surface materials:

1. Conductive adhesives should be used to bond fused silica, Kapton, and Teflon second-surface mirrors to conductive substrates that are grounded to structure. If the substrate is not conductive, metal foil or wire ground links should be laminated in the adhesive and bolted to structure. Only optical solar reflectors (OSRs) with conductive back surfaces (e.g., Inconel should be used.
2. When conductive adhesives are used, the long-term stability of the materials system must be verified, particularly conductivity in vacuum after thermal cycling, compatibility of the materials (especially for epoxy adhesive) in differential thermal expansion, and long-term resistance to galvanic corrosion.
3. Metallized Teflon® is particularly susceptible to ESD degradation, even when grounded. Avoid using it. If there is no substitute for a specific application, the effects of EMI, contamination, and optical and mechanical degradation must be evaluated.
4. Paints (ESD-conductive/leaky) should be applied to grounded, conductive substrates; the primer must be conductive, too. If painting over a grounded surface is not possible, paint coverage should be extended to overlap grounded conductors around the paint's perimeter.
5. Ground tabs must be provided for freestanding (not bonded down) dielectric films with conductive surfaces.
6. Meshes that are simply stretched over dielectric surfaces are not effective; they must be bonded or heat-sealed in a manner that will not degrade or contaminate the surface.
7. There are several techniques for grounding thin, conductive front-surface coatings such as ITO. At least one commercial manufacturer has found that the added cost of a reliable ITO coating and grounding/referencing method on OSRs and cover glasses has provided excellent in-orbit performance and thus is worth that cost. The methods include welding of ground wires to front-surface metal welding contacts, front-surface bonding of coiled ground wires (to allow for differential thermal expansion) by using a conductive adhesive, and chamfering the edges of OSRs before ITO coating to permit contact between the coating and the conductive adhesive used to bond the OSR to its substrate.
8. For MLI, extending the aluminum foil tab to the front surface is suitable.

3.2.1.5.2 Nonconductive Surfaces. If the spacecraft surface cannot be made 100% conductive, an analysis must be performed to show that the design is acceptable from an ESD standpoint. Note that not all dielectric materials have the same charging or ESD characteristics. The choice of dielectric materials can affect surface voltage profiles significantly. For example, it has been shown [3,4] that different cover slide materials have differing resistivities and that all are affected by temperature. Cover slide material can noticeably affect spacecraft charging.

An adequate analysis preceding the selection of materials must include a spacecraft charging analysis to determine surface potentials and voltage gradients, spark-discharge parameters (amplitude, duration, frequency content), and EMI coupling. The cost and weight involved in providing adequate protection (by shielding and electrical redesign) could tilt the balance of the trade-off to favor the selection of less optically transmissive cover slides that are more reliable from spacecraft charging, discharging, and EMI points of view.

The proven materials have their own cost, weight, availability, variability, and fabrication effects. In addition, uncertainties relating to spacecraft charging effects must be given adequate consideration. Flight data have shown apparent optical degradation of standard, stable thermal control materials (e.g., OSRs and Teflon second-surface mirrors) that is in excess of ground test predictions, part of which could be the result of charge-enhanced attraction of charged contaminants. In addition, certain spacecraft anomalies and failures may have been reduced or avoided by using charge-control materials. When the spacecraft design is completed, the remaining dielectric materials on the surface of the spacecraft must be evaluated for their ESD hazard. Evaluate potential stored energy and nearby potential victims to see if a spacecraft threat exists.

A spacecraft with larger portions of dielectric may have retarding electric fields because the dielectric diminishes the effects of the photoemission process [5]. As a result, the spacecraft structure potential may go more negative and thus reduce the differential voltage between the dielectric and the spacecraft. The lesson to be learned is that all surface dielectrics must be examined for their differential charging. Each dielectric region must be assessed for its breakdown voltage, its ability to store energy, and the effects it can have on neighboring electronics (disruption or damage) and surfaces (erosion or contamination).

3.2.1.5.3 Surface Secondary Emission Ratios. Other means of reducing surface charging exist, but they are not well developed and are not in common use. One suggestion for metallic surfaces is an oxide coating with a high secondary electron yield. This concept, in a 3-D surface charging simulation, reduced charging of a spacecraft dramatically and reduced differential charging of shaded Kapton slightly. Any materials selected should be analyzed carefully to ensure that they do not create problems of their own and will work as intended over their service lives.

3.2.1.6 Radiation Spot Shields and Other Floating Metals. Grounding radiation spot shields is essential (i.e., radiation spot shields must be grounded). Bodeau [6,7], in particular, emphasizes this rule. Grounding can be done in a number of ways. If a Solithane or other conformal coating has adequate resistivity (on the order of $10^{10}\ \Omega \cdot \text{cm}$ or less), a separate ground wire is unnecessary. It must be determined that any solution, such as partially conductive Solithane, will not degrade (increase resistivity) in the expected radiation and long-term vacuum environments. (This relatively large resistivity, $<10^{10}\ \Omega \cdot \text{cm}$, is generally acceptable on an interior surface since charging fluxes are lower on the interior of a spacecraft. Check actual charging fluxes if uncertain about a particular application.)

3.2.1.7 Filter Circuits with Lumped Elements or Circuit Choices. Use low-pass filters on interface circuits. Use low-speed, noise-immune logic, if possible. Use complementary metal-oxide-semiconductor (CMOS) circuits that have higher interface noise immunity. Beware, however, of the latch-up sensitivity of CMOS. For IESD purposes, the filter or protection network must be applied so that it is physically at the device terminals.

Electrical filtering should be used to protect circuits from discharge-induced upsets. All circuits routed into the Faraday cage region, even though their wiring is in shielded cabling, run a greater risk of having ESD-caused transient voltages on them. Initial design planning should include ESD protection for these circuits. It is recommended that filtering be applied to these circuits unless analysis shows that it is not needed.

The usual criterion suggested for filtering is to eliminate noise shorter than a specific time duration (i.e., above a specific frequency). On the Communications Technology Satellite (CTS), in-line transmitters and receivers effectively eliminated noise pulses of less than 5-μs duration, which were suitable to its circuitry. Similar filtering concepts might include a voltage threshold or energy threshold. Filtering is believed to be an effective means of preventing circuit disruption and should be included in system designs. Any chosen filtering method should have analyses and tests to validate the selected criteria. Filters should be rated to withstand the peak transient voltages over the mission life. Today's circuitry with smaller feature sizes and lower operating voltages may need even more stringent filtering for ESD protection.

3.2.1.8 Isolate Transformer Primary-to-Secondary Windings. Isolate the primary and secondary windings of all transformers. Reduce primary-to-secondary winding capacitance to reduce common-mode noise coupling. This is an EMC solution to reduce coupling of ESD-induced noise.

3.2.1.9 Bleed Paths for Forgotten Floating Conductors. Provide a conductive bleed path for all conductors (including structural elements), including, but not to be limited to, the following items:

- Signal and power transformer cores
- Capacitor cans
- Metallic IC and hybrid device cases
- Unused connector pins and unused wires in cables, including those isolated by switching
- Relay cans

These items may be protected by stray leakage by deliberate resistors to ground, through their conformal coating, normal bleed paths, or small charge/energy storage areas. Ensure that the presumed bleed path really works or that the ungrounded items are not an ESD threat before depending on stray leakage for ESD protection.

3.2.1.10 Interior Paints and Conformal Coatings. Most paints and conformal coatings are dielectrics and can be charged by energetic particles. This must be considered in evaluating the likelihood of interior charging of a design. If conductive coatings are used, these must be grounded to the structure to allow charge to bleed off. For conductive coatings, conductive primers must be used. If nonconductive primers are used, the conductive coating will be isolated from ground and will charge. Other grounding means must be provided if the primer or substrate is nonconductive.

3.2.1.11 Cable Harness Layout. Route cable harnesses away from apertures. Care should be taken in the layout of the internal electrical harnesses to minimize exposure to the environment's energetic particles. The harness should not be close to the edges of apertures.

3.2.1.12 External Wiring. Provide additional protection for external cabling. Cables external to the spacecraft structure should be given adequate protection. The dielectric coatings can charge to a point where discharge can occur. At present, there are no simple design rules for the degree of shielding needed. Cables should be tightly wrapped to minimize gaps where discharges can propagate.

3.2.1.13 Slip Ring Grounding Paths. Carry bonds and grounds across all articulated and rotating joints. For a rotating joint with slip rings, the chassis or frame ground (bond) must be carried through the slip ring also and then grounded. Note that for the case of the solar array and other situations that may involve transfer of ESD current, a series resistance in the path from spacecraft frame to solar array frame will be required to limit the amount of current that can carry this ESD current into the satellite (Section 3.2.4.3, rule 20).

3.2.1.14 Wire Separation. Segregate cabling from outside the spacecraft after it enters the Faraday cage. Wires coming from outside the spacecraft should be filtered, preferably at the entry point but certainly before being routed with other interior cabling. This is based on an assumption of external ESD noises and is to prevent coupling to the interior. It is a poor design practice to route the filtered and unfiltered wires together in the same bundle because noise can be coupled between them.

3.2.1.15 ESD-Sensitive Parts. Pay special attention to ESD-sensitive parts. In the parts list, flag all parts that are class 1 ESD-sensitive in accordance with MIL-STD-883G, Test Method Standard for Microcircuits [8] [Method 3015.7,

Electrostatic Discharge Sensitivity Classification (Human Body Model)]. Do a charging analysis after completion of the spacecraft design. Evaluate the charging rates with respect to Section 3.2.2 parameters. Protect the devices if they might be damaged by an expected threat.

3.2.1.16 ***Procedures.*** Institute proper handling, assembly, inspection, and test procedures to ensure the electrical continuity of the space vehicle grounding system. The continuity of the space vehicle electrical grounding and bonding system is of great importance to the overall design susceptibility to spacecraft charging effects. In addition, it will strongly affect the integrity of the space vehicle EMC design. Proper handling and assembly procedures must be followed during fabrication of the electrical grounding system. All ground ties should be inspected carefully, and dc resistance levels should be tested during fabrication and again before delivery of the space vehicle. A final check of the ground system continuity during preparation for space vehicle launch is desirable. A related reference is NASA-HDBK-4001 [1], which describes how to establish an electrical grounding architecture system for power and signals. This design book is complementary to the ESD effort.

3.2.2 Surface ESD Design Guidelines, Excluding Solar Arrays

3.2.2.1 ***Qualitative Surface ESD Guidelines.*** Refer to the General ESD Design Guidelines, Section 3.2.1.

3.2.2.2 ***Quantitative Surface ESD Guidelines.*** These detailed surface conductivity design guidelines are equation-based to assist designers in accounting for differing geometries and material conductivities. Since these are general, projects may formulate their own rules.

To discharge surfaces that are being charged by space plasmas, a high resistivity to ground can be tolerated because the plasma charging currents are small. The following guidelines are suggested:

1. Conductive materials (e.g., metals) must be grounded to structure with resistance, expressed in ohms:

$$R < \frac{10^9}{A} \tag{3.2-1}$$

   where A is the exposed surface area of the conductor (cm^2)

2. Partially conductive surfaces (e.g., paints) applied over a grounded conductive surface must have a resistivity–thickness product, expressed in $\Omega \cdot \text{cm}^2$:

$$rt < 2 \times 10^9 \tag{3.2-2}$$

   where r is the material resistivity (Ω-cm) and t is the material thickness (cm).

3. Partially conductive surfaces applied over a dielectric and grounded at the edges must have material resistivity, expressed in $\Omega \cdot \mathrm{cm}$, such that

$$\frac{rh^2}{t} < 4 \times 10^9 \qquad (3.2\text{-}3)$$

where h is the greatest distance on a surface to a ground point (cm).

The guidelines above depend on the particular geometry and application. A simplified set of guidelines is supplied for early design activities as follows:

1. Isolated conductors must be grounded with less than $10^6\ \Omega$ to structure. This is the same value as that recommended in ECSS-E-ST-20-06C [9].
2. Materials applied over a conductive substrate must have bulk resistivities of less than $10^{11}\ \Omega \cdot \mathrm{cm}$.
3. Materials applied over a dielectric area must be grounded at the edges and must have a resistivity less than $10^9\ \Omega$ per square. The term Ω *per square* is defined as the resistance of a flat sheet of the material measured from one edge of a square section to the opposite edge. (This is described more fully in Section E.3.)

These requirements are more strict than the preceding relations, which include the effects of spacecraft geometry. In all cases, the usage or application process must be verified by measuring the resistance from any point on the material surface to structure. Problems can occur. For example, one case was observed where a nonconductive primer was applied underneath a conductive paint; the paint's conductivity was useless over the insulating primer.

All grounding methods must be demonstrated to be acceptable over the service life of the spacecraft. It is recommended that all joint resistances and surface resistivities be measured to verify compliance with these guidelines. Test voltages to measure resistivity of dielectric samples should be at least 500 V. See Section E.4 for measurement examples.

Grounding methods must be able to handle current bleed-off from ESD events, vacuum exposure, and thermal expansion and contraction. As an example, painting around a zero-radius edge or at a seam between two dissimilar materials could lead to cracking and a loss of electrical continuity at that location.

3.2.3 Internal ESD Design Guidelines

Guidelines for internal hardware are often the same as for the guidelines for surfaces.

3.2.3.1 Qualitative Internal ESD Guidelines. Refer to the General ESD Design Guidelines, Section 3.2.1. Internal regions also have surfaces, and surface rules apply.

3.2.3.2 Quantitative Internal ESD Guidelines. Quantitative guidelines are recommended in the following sections.

3.2.3.2.1 Grounding Conductive Elements. Unused spacecraft cables, circuit traces, and other noncircuit conductive elements greater than 3 cm^2 in surface area (0.3 cm^2 for conductive elements on circuit boards) or longer than 25 cm in length must be ground referenced; be sure to provide a deliberate or known bleed path must for all radiation spot shields. For other wires and metal, being in a circuit is usually adequate. It is best not to have any deliberate ungrounded metals, including unused connector pins as an example. Exceptions are allowed in situations in which one of the following conditions is true:

1. Discharges will not occur in the expected charging environment.
2. The discharges expected to occur will not damage or disrupt the most sensitive circuits in the vicinity or cause EMI that exceeds the EMC requirements, assuming that separate EMC requirements exist.

These historic quantitative guidelines may need reconsideration for newer spacecraft. For example, ECSS-E-ST-20-06C [9] recommends a maximum of 1 cm^2 ungrounded metal on the surface of a spacecraft.

3.2.3.2.2 Shielding to Limit Internal Electron Fluxes. Determine electron fluxes at all part locations using a worst-case electron spectrum (Fig. 2.6 for GEO), and shield all electronic circuitry to the following levels (Fig. 2.5 basis with no margin; projects may wish to consider margins). GEO orbit approximate rule of thumb to limit IESD: If there are 110 mils of aluminum-equivalent shielding, as stated previously, there is no need to shield further and there is no need to do an electron transport analysis unless there is a desire to save weight (GEO orbit approximate rule only). Bodeau's [6,7] recommendations for lower flux limits have the effect of raising this to 200+ mils of aluminum shielding in Earth GEO orbits.

If the flux computed at the location is less than 0.1 pA/cm^2, the circuit needs no additional shielding (any electron environment). (Basis: less than 2×10^{10} e/cm^2 deposited in 10 h—using only the incident fluence is more conservative.) Note, however, that Bodeau [6,7] and Balcewicz et al. [10] recommend one-tenth of this (0.01 pA/cm^2), which begins to present difficulties in implementation. Note also that this recommendation depends on the assumed room-temperature bulk resistivities of commonly used dielectric materials. For applications that are constantly at cryogenic temperatures, the flux limit must be adjusted downward to account for the increased cryogenic bulk material resistivities (see, in particular, Bodeau [7]). Also note that ECSS-E-ST-20-06C [9] and Bodeau [7] recommend longer flux integration times to account for dielectric materials with time constants greater than 10 h.

If the incident flux is between 0.1 and 0.3 pA/cm^2, shield to a level of 0.1 pA/cm^2 if the circuitry is Class 1 ESD-sensitive (MIL-STD-88G3 [8], Method

3015.7) or if this type of circuitry has had a known on-orbit anomaly. (Again, remember that Bodeau indicates that less flux/more shielding may be appropriate for very sensitive circuits.) If the incident flux is between 0.3 and 1 pA/cm^2 and Class 2 ESD-sensitive or greater circuitry is present, then shield to <0.3 pA/cm^2. If the incident flux is greater than 1 pA/cm^2, IESD problems may exist.

3.2.3.2.3 Filter Circuits. For wiring protected less than the levels of Section 3.2.3.2.2, protect attached circuits by filtering. To protect the interior sensing circuit for temperature transducers that are located outside the main box of the spacecraft, resistor–capacitor (RC) filters or diode protection can be used to suppress any ESD effects. Another reason for filtering is if the shielding levels of Section 3.2.3.2.2 cannot be achieved. The filter should anticipate a pulse on the order of 20 ns wide. As a rough example, filtering should protect against a 20-pF capacitance charged with 100 nC (about 5 kV stress, 250 μJ). The real estimated threat should be used, if possible.

3.2.3.2.4 Voltage Stress. Keep the electric field stress in dielectrics below 100 V/mil ($\sim 4 \times 10^4$ V/cm or 4×10^6 V/m; see [11,12]). When designing high-voltage systems, keep the electric field below 100 V/mil in any material or gap. This voltage stress could be in circuit board dielectrics being charged by the incident electron flux while the adjacent metals remain at a low voltage. Other sites of concern are ungrounded metal radiation shields on insulating surfaces charged by the electron flux while the adjacent surfaces remain at low voltages or insulated surfaces being charged while internal wires remain at low voltage. Power supplies can sustain a discharge after an arc has been initiated, so power wiring should never be bare (exposed). All such possible sources must be eliminated where possible.

3.2.3.2.5 Coat Circuit Boards with Leaky Dielectric. Use leaky/conductive conformal coating on circuit boards. Leung and Mikkelson [13] use a $10^{10}\ \Omega \cdot \text{cm}$ clear coating, resulting in an automatic bleed path of resistance (R) such that $10^9\ \Omega < R < 10^{13}\ \Omega$. This shunt leakage will not affect circuit operation but can bleed off most levels of internal charging. The coating has been space-qualified (temperature cycle, vacuum, etc.). It has demonstrated dramatic reduction in discharge voltages on its victims in laboratory tests and does not involve circuit or board layout changes. At present, the specific formulation is a proprietary product, but the concept could be adapted.

3.2.3.2.6 Fill Circuit Board Material with Grounded/Referenced Metal. Limit the regions where charge can accumulate. Place grounded (best) or referenced traces in open (unused) areas. This is a new idea in this book (based on NASA-HDBK-4002A) to minimize the size of any ESD arc inside a circuit board by reducing the dielectric volume that might contain a discrete lump of ESD energy. It was

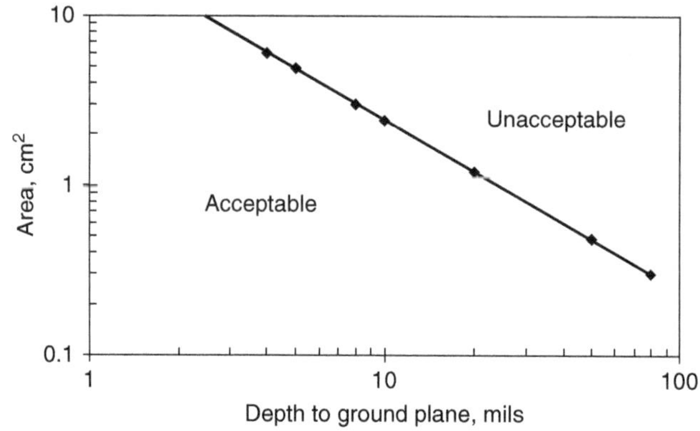

Fig. 3-1. Permissible area versus depth-to-ground plane.

not developed in response to a specifically identified failure in space and *has not been validated*. The derivation is shown in Appendix H.

Circuit boards should be designed so that any metal area greater than 0.3 cm^2 should also have a bleed path with the same ESD grounding limits of 0 to 10 MΩ resistance to ground. Circuit boards should be designed so that there will be no open (unused) surface areas greater than 0.3 cm^2. Otherwise, place a metal land that is ESD grounded with 0 to 10 MΩ resistance to ground in the unused dielectric area.

This effect is shown in Fig. 3-1, which also proposes a new rule for circuit board exposed dielectric areas. [The term *ground* in Fig. 3-1 means (a) not floating or (b) referenced within the circuit.] The design rule assumes a standard FR4 circuit board material of 80-mil thickness. The term *depth to ground plane* means the distance from any dielectric to a ground-referenced plane. For example, if the board is 80 mils thick with a ground plane on one external surface, the depth to ground plane is 80 mils; if both exterior surfaces are ground (or power) planes, the depth to ground plane is 40 mils.

3.2.4 Solar Array ESD Design Guidelines

This section contains guidelines to protect solar arrays from ESD charging problems.

3.2.4.1 Solar Array Possible ESD Problem Areas. Solar arrays, with their possibly high operating voltages and their available power, can cause the following spacecraft charging effects:

1. Arcing with loss of power and permanent damage to the solar arrays if the arcing is sustained by power from the array itself or by power from the spacecraft internal stored energy.
2. Arcing with momentary loss of power and degradation of solar arrays (similar to the preceding but without a sustained arc).

3. Charging of spacecraft structures with respect to the plasma and resulting problems (contamination by attraction of charged surfaces and/or possible erosion of surfaces as species are attracted to the surface). This is very noticeable in LEO environments [14].
4. Disruption of science (electric fields from the surface potentials of solar arrays will alter the path of electrons and ions so that plasma measuring instruments will not record the proper directionality of electrons and ions entering their field of view).
5. Loss of power related to current leakage at exposed conductors in the array. A dramatic rise in power loss can occur at string potentials of about 200 to 1000 V positive with respect to plasma potential related to the phenomenon of snapover. At a geometry- and material-dependent voltage, the current in the array's current–voltage (I/V) curve makes a dramatic change to increasingly larger currents because of enhanced secondary emission and greater plasma contact area.

3.2.4.2 ***Background.*** The following are basic rules to avoid spacecraft charging issues related to solar arrays that can cause surface damage, upset science instruments on the spacecraft, or may result in power loss to the space plasma, and resulting ESDs and damage. The rules are gleaned from several sources. Good references for this subject include references [3,14–17] and references therein.

Note that there has not been enough flight experience with higher voltage solar arrays (operating voltages greater than 28 V) at the time of this writing (2011) to generate guaranteed and optimal design rules for any space plasma environment situation, so the following should only be considered as guidelines. (*Exception*: The ISS uses higher voltage and massive arrays. Their impact has been investigated in several papers [18].) The principal rule still must be: Test any new design in the anticipated environment. There has been considerable focus on solar arrays [19], and there will be more in the next few years. Progress, especially in the design of test protocols that expand on the existing ideas and rules contained herein, is anticipated in the near future.

It is not necessary to use all the design ideas listed herein because that would cause excess mass, excess cost, reduced efficiency, and so on. Trade-offs are needed to achieve an adequate design. The point is that after the design has been optimized by engineering and analysis, the final design must be verified by test with as realistic test conditions as possible. The test considerations are described in the following material with a shopping list of design features. To illustrate the severity of the problem, Fig. 3-2 shows the type of damage that may occur to solar arrays if the design is inadequate in a space plasma environment.

Figure 3-2a is a photograph of a solar array recovered from the European Space Agency European Retrievable Carrier mission by the Space Shuttle. As space failures typically are not retrieved, ground tests have to be performed for failure analysis. These do not, however, represent an actual product that failed in space as Fig. 3-2a shows. As an example of the corresponding ground simulation,

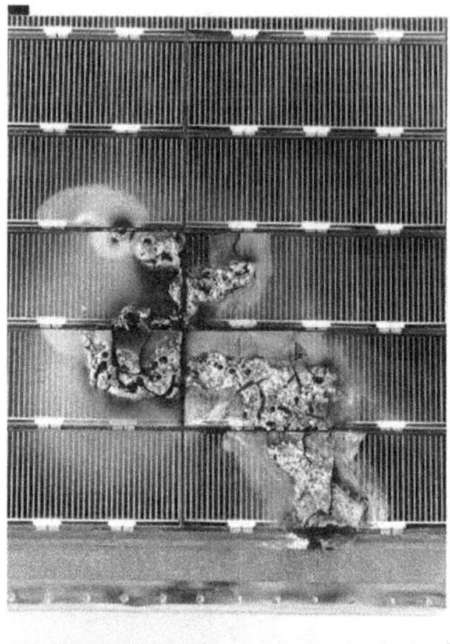

(a) Failure caused by in-flight arcing.

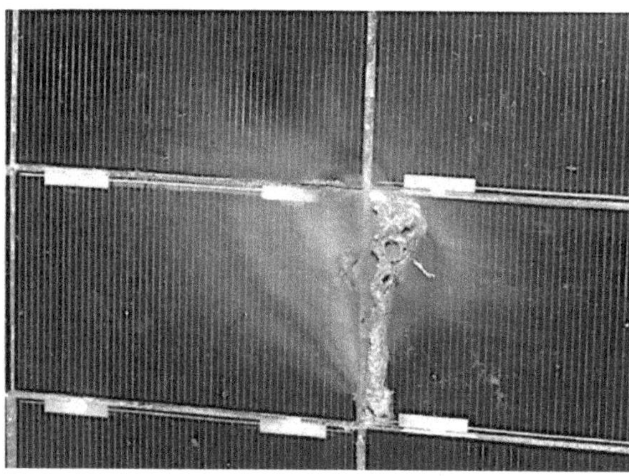

(b) Failure caused by ground ESD arcing.

Fig. 3-2. Examples of solar array failures caused by (a) in-flight arcing (from [15]) and (b) ground ESD arcing. (*See insert for color representation of the figure.*)

Fig. 3-2b shows a solar array that failed in a plasma environment during a ground test [20,27].

3.2.4.3 Solar Array Design Guidelines to Protect Against Space Charging and ESDs.

1. Build solar arrays so they do not arc. This is a difficult requirement with the present trend toward higher power solar arrays with higher voltages (to minimize wiring size and weight).

2. Test any new design in a representative plasma and energetic particle environment; test with a voltage margin on the solar array to assure that the design is adequate.
3. Arrays with 40 V or less maximum cell-to-cell potential difference are assumed not to be a hazard with margin. This has been measured to be a reasonable guideline. Potentials on the order of 80-V cell-to-cell potential difference can, however, initiate arcs on unprotected solar array designs. Note that string voltages might be about 20% higher than nominal if they are not carrying current/open-circuited.
4. Place diodes in series with each string so that an arc on a single string will not be sustained by energy/current from the other strings on the array or the main bus stored energy. Available currents on the order of 2 A can sustain an arc with unprotected solar array designs. Size the diodes to tolerate the maximum anticipated ESD arc or short circuits to chassis. Figures 3-3 and 3-4 illustrate rules 3 and 4.
5. Especially for LEO, consider building arrays so that they are not grounded negatively to the spacecraft frame/chassis ground. If the power converter can provide isolation, a *floating array* is one option. With this design, the array voltage with respect to the plasma will adjust to minimize power loss currents from the array through the plasma potential (assuming a conventionally built array, with exposed cell potentials on the edges). This results in a (soft) virtual ground such that about 5% of the array area is higher than the plasma potential, and 95% of the array is lower than the plasma potential. The authors generally oppose any totally floating conductor system.

   An alternative option to floating that addresses the same issue is to ground the solar array at the positive end. This has less effect on the overall spacecraft potential and less current/energy losses to space. The best fixed grounding solution to keep the spacecraft frame at plasma potential is to ground the solar array strings to frame at about 5 to 10% of the distance (potential) from the positive end of the solar array. The objectives are to reduce the power loss of leakage current through the plasma and to reduce the voltage of any one part of the array with respect to local plasma below potentials that could trigger an arc. The two objectives do not have the same solution, so a compromise may be necessary. Analyses of the applicable charging currents, power loss, and resulting voltage balance should be done before adopting this design approach [15]. The reason that this design might be more useful at LEO is that the greater plasma density has a greater impact on the space charging concerns listed in these rules. A similar situation may exist if an electric thruster effluent impacts the solar array or if some other higher-density plasma surrounds the arrays.
6. Design the solar arrays to avoid excessive power loss (e.g., keep the positive voltage with respect to frame less than about 100 V). The remedy here if high voltages must be used is to insulate the high-voltage metal regions (interconnects and wiring) with insulating grout [space-qualified

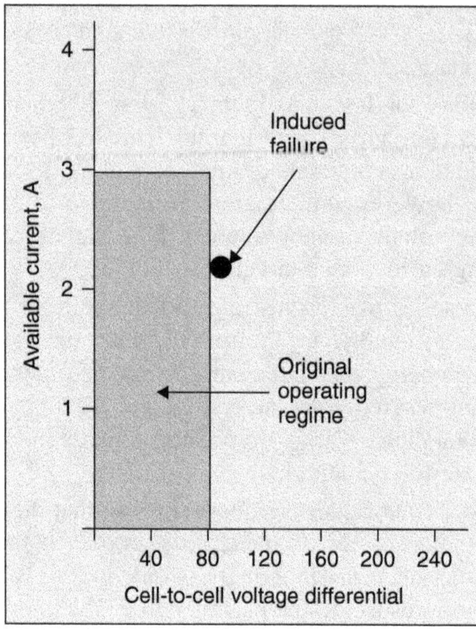

Fig. 3-3. Measured gallium arsenide (GaAs) coupon I/V failure threshold (From [3].) (*See insert for color representation of the figure.*)

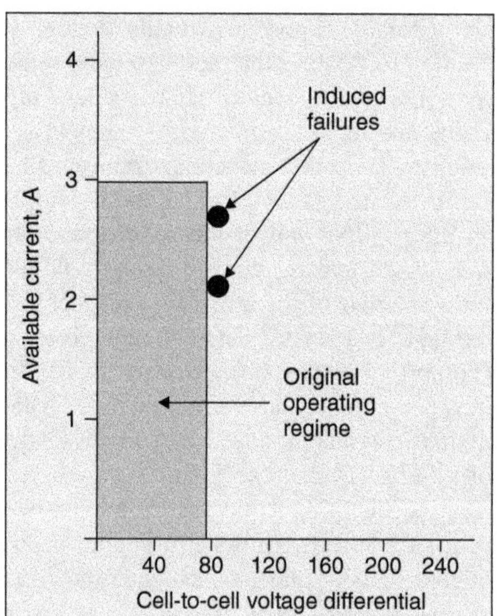

Fig. 3-4. Measured silicon (Si) coupon I/V failure threshold (From [3].) (*See insert for color representation of the figure.*)

room-temperature vulcanized (RTV) silicone]. Avoid any air pockets or voids in the grouting. The latter instruction is very important because entrained air can assist in creating a Paschen discharge, meaning that it takes less voltage to trigger an arc. The fabrication processes must be well thought out, the assembly personnel must be well trained, and fabrication inspections (quality assurance) must be part of the process.

7. Do not vent any gas onto or in the vicinity of exposed solar array potentials. A discharge can be triggered at lower voltages in the presence of the resulting partial pressure regimes. Most typically, the gas would be attitude-control gas venting but could also be cryogenic cooler gas venting (again, a possible Paschen discharge).
8. Insulate the solar arrays so that there is no potential-carrying conductor exposed to space. The simplest concept is to grout all the spaces between solar cells as in rule 6 in this section. Figures 3-5 and 3-6 illustrate the configuration being discussed. Figure 3-6 illustrates a shortcut that may be permissible if testing demonstrates its adequacy. Figure 3-6 assumes that cells 1 and 3 are connected in a string and that the potential between them is small, so no grouting is placed between them. Cells 1 and 2 and 3 and 4, by contrast, are adjacent strings with different potentials and need insulation the full distance of their shared edge. At the regions labeled "RTV Barrier", the RTV is extended out a bit at the corner as an extra insulation where higher electric fields may be present. In Fig. 3-6, b is grout width; in Fig. 3-5, b, r, g, and x are variables used in equations from reference [3]. A full RTV barrier would be the most robust design. Figures 3-7 and 3-8 show the improvement when grouting is used compared to the original operating regime illustrated in Figs. 3-3 and 3-4.
9. Use slightly conductive cover slides to limit electric fields at potential arc sites.
10. Use cover slides with large overhang to limit electric fields in the plasma region.
11. Limit the differential potential between adjacent cells in the array to reduce arc likelihood. As a limit, 40 V is suggested, but test the array design.
12. Make the cell intergap spacing wide enough so that there will be no arcing. Testing in plasma must be performed for the candidate designs chosen. This design solution is less likely, because it reduces the cell density and thus results in less power density (W/m^2 and/or W/kg).
13. Verify that solar array materials will not outgas in space or decay at high temperatures.
14. Make all insulating materials thick enough to withstand the anticipated electric fields so that they are below the breakdown voltage of that material. Do not make the materials so thick that they accumulate charge to the degree that they cause problems. Make this part of the ESD analysis/test process.

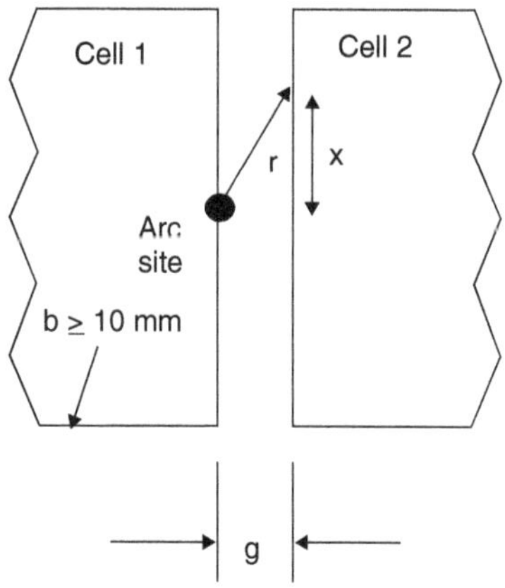

Fig. 3-5. Intercell gap. (From [3].)

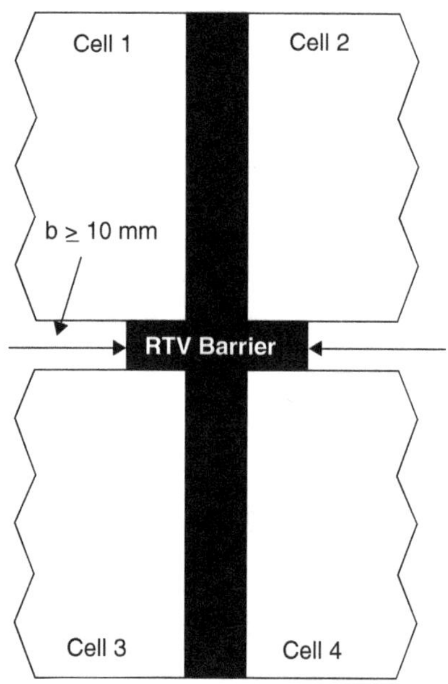

Fig. 3-6. Grouting barrier to stop arcs. (From [3].)

15. Use a plasma contactor (neutral plasma beam) on the spacecraft as a means to keep the spacecraft at the plasma potential. This is useful in LEO or for performing low-energy plasma measurements or to reduce erosion of surfaces caused by impact of attracted charged particles. Any such active device carries reliability concerns in addition to weight, complexity, power consumption, and consumables, but the ISS contactors are working well.
16. Use thin dielectrics with resistivities such that a charge will not build up in the anticipated environment. Examples include wire insulation, substrates,

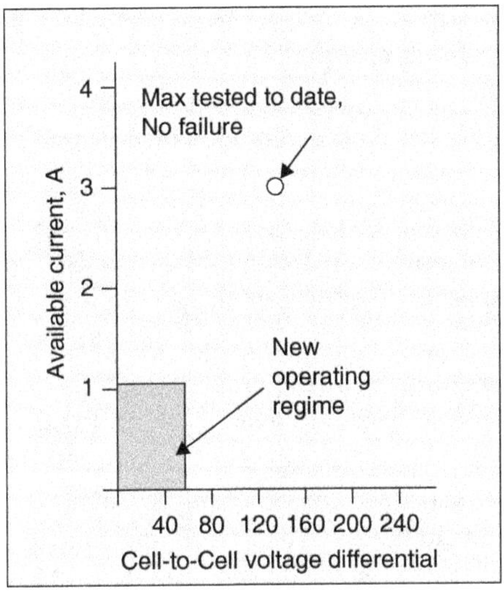

Fig. 3-7. GaAs coupon with RTV barrier installed. (From [3].)

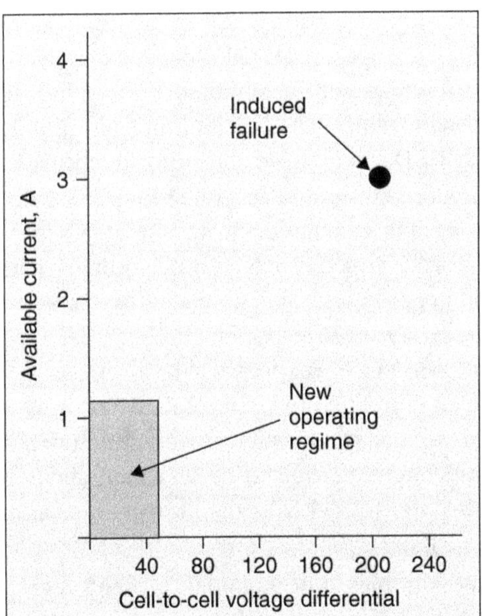

Fig. 3-8. Si coupon with RTV barrier installed. (From [3].)

and structures. The idea is to make the resistivity/thickness combination so that charge can bleed off through the material to ground faster than hazardous potentials can arise on the material surface or in its volume.

17. Do not put ESD-sensitive electronics near where a solar array discharge may occur. An example would be a thermistor or its wiring placed near

the solar cells so that ESD energy can be carried back to an ESD-sensitive telemetry data multiplexing unit.

18. Filter solar array wiring, preferably at the entry to the spacecraft Faraday cage, but definitely before it enters the power supply. If solar array wiring is not filtered at the entry point to the Faraday cage, shield the wiring from that point to the power supply.
19. Filter temperature sensors and other data signals from the solar array as they enter the spacecraft or at least at the entry point into their electronic sensing box.
20. Isolate the solar array substrate ground from spacecraft chassis ground. Place an isolation resistance of about 2 to 250 kΩ between the solar array substrate/frame and the spacecraft chassis. This will limit currents from the solar array to its substrate and returning through the spacecraft structure. The resistance should be calculated for all the parameters of the solar array and environment. This is a new rule compared to NASA TP-2361 [5], which had recommended that the solar array structure be carefully grounded to the spacecraft structure. Extra mechanical complexity will be required to provide the necessary insulation between the main spacecraft and the solar array structure. See Bogus et al. [21] for an example.

    The resistor lower-bound size should be a value that limits any fault currents to a small value that will interrupt any holding currents caused by a triggering ESD event from the array to the structure. Assuming 1 mA as a maximum permissible sustained fault current (very conservative) on a 100-V array, the calculation would be 100 V/mA or 100 kΩ as the minimum solar array structure isolation from the spacecraft chassis.

    The resistor upper-bound sizing relates to controlling the differential potential of the array with respect to the chassis. For example, if space plasma charging currents are expected to be 1 nA/cm^2 (GEO), the maximum value of collected current would be calculated as array area $\times 1$ nA/cm^2. If we assume that the maximum array support structure potential with respect to the spacecraft bus is desired to be less than about 10 V and the array area is 4 m^2, this gives 250 kΩ as the maximum solar array isolation from the spacecraft chassis.
21. Consider the possibilities. For example, in LEO regimes, the plasma can initiate an arc for 75-V arrays, and the arc can be sustained by the power of the solar array. At GEO and other locations, the arc initiator could be charging of the dielectric surfaces (this environment requires perhaps as much as a 400-V differential to the array wiring) in the vicinity of a conductor with the same result. The design should accommodate any situation that occurs, with focus on the anticipated environment, if known. Extreme temperatures, solar illumination, cell-to-cell potentials, and plasma density and temperatures are some of the environmental parameters.

22. Consider Si cells versus GaAs cells. It may be that Si or GaAs cells are inherently less likely to have ESDs. To date, the data have too many variables to say which is better, but future research may determine that there is an advantage to one or the other.
23. Insulate the solar array connector wires leading into the spacecraft as much as possible. Solar array drive assembly details include isolating wiper arms and slip ring spacer insulator height [22].
24. Consider use of the stretched lens array as advocated by Brandhorst et al. [23]. This is a concentrator technology that may eliminate many space charging problems with solar arrays and has been space qualified.

3.2.4.4 Solar Array Testing Rules for Space Charging Characterization. Figure 3-9 shows the typical elements of a solar array ESD charging threat test. Figure 3-9 is intended to provide a simple introduction to test needs. Many solar array test plans become increasingly complex with attempts to add better simulation of reality but in a limited test space and with sample coupons rather than the real full-size article. The test layout in Fig. 3-9 may

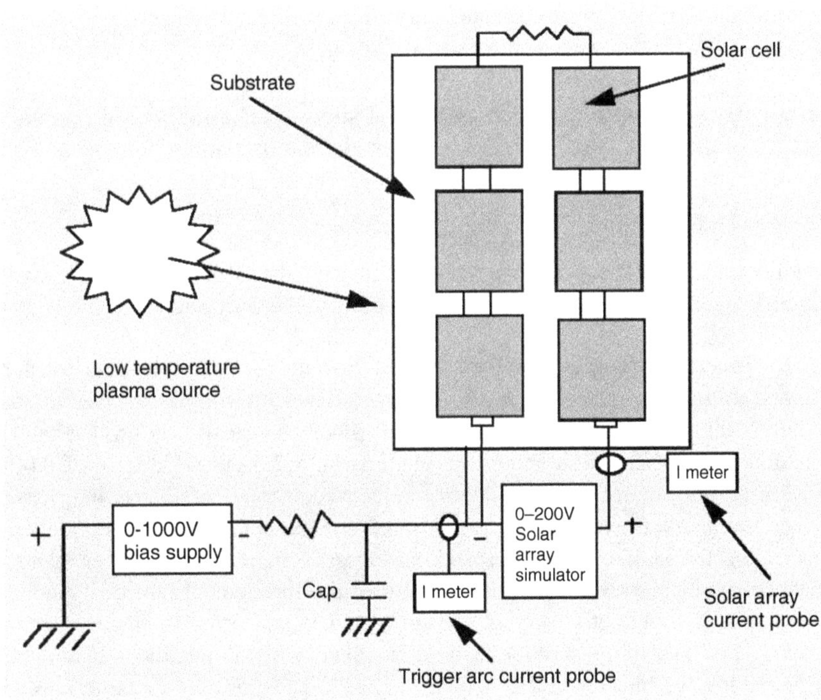

Fig. 3-9. NASA Lewis Research Center (now Glenn Research Center) solar array space charging and ESD test setup. (From [3].)

be modified to reflect a more specific knowledge of solar array equivalent schematics or changed if newer applicable requirements documents become available. Additional details involve capacitances to simulate stored energy in the capacitance of the cells that can cause an initial high current pulse, inductances in wiring that can cause ringing and resonances, and a grounded substrate that may provide a ground return for an arc. A well-thought-out test has a number of details needed to simulate the space situation as closely as possible.

Test parameters that add to the complexity include:

- Actual spacing and construction of the solar array
- Simulation of the plasma environment
- Simulation of the higher-energy electron environment
- Simulation of the Sun
- Temperature of the array/cover glass, including occultation (no solar simulation)
- Energy storage of a string (capacitance to ground, if a partial array is used)
- Simulation of the solar array dynamics, including transient voltage slew rate and capacitance to ground
- Simulation of the wiring (capacitance and inductance effects)
- Presence of grounded or isolated cell substrate

In an excellent paper, Amorim et al. [24] show solar array arcing current as measured in the laboratory, with a discussion and interpretations for space needs.

3.2.5 Special Situations ESD Design Guidelines

The guidelines in this section are special situations that are treated more easily separately. General ESD design guidelines are provided in Section 3.2.1.

3.2.5.1 Thermal Blankets. All metallized surfaces in MLI blankets must be grounded to the structure electrically. The metallized multilayer surfaces in each separate blanket should be grounded to each other electrically by ground tabs at the blanket edges. Each tab should be made from a 2.5-cm-wide strip of 0.005-cm-thick aluminum foil. The strip should be accordion folded and interleaved between the blanket layers to give a 2.5 × 2.5 cm contact area with all metallized surfaces and the blanket front and back surfaces. Nonconductive spacer or mesh material must be removed from the vicinity of the interleaved tab; or it must be verified that all conductive layers are grounded if spacer/mesh material is not removed. The assembly should be held in place with a metallic nut and bolt that penetrates all blanket layers and captures 2.0-cm-diameter metallic washers positioned on the blanket front and back surfaces and centered in the 2.5 × 2.5 cm tab area. The washers may have different diameters, with the inner surface of the smaller washer recessed to ensure maximum peripheral contact area between the interleaved foil strip and each metallized blanket surface. The tab should be grounded to structure by a proven technique such as a wire that is as short as possible (15 cm maximum) or conductive Velcro.

Redundant grounding tabs on all blankets should be implemented as a minimum. Tabs should be located on blanket edges and spaced to minimize the maximum distance from any point on the blanket to the nearest tab. Extra tabs may be needed on odd-shaped blankets to meet the condition that any point on a blanket should be within 1 m of a ground tab.

The following practices should be observed during blanket design, fabrication, handling, installation, and inspection:

1. Verify layer-to-layer blanket grounding during fabrication with an ohmmeter.
2. After installation, verify less than 10-Ω dc resistance between blanket and structure with an ohmmeter. (Verification details are in the test procedures.)
3. Close the blanket edges (cover, fold in, or tape) to prevent direct irradiation of inner layers.
4. Do not use crinkled, wrinkled, or creased metallized film material.
5. Handle blankets carefully to avoid creasing of the film or possible degradation of the ground tabs.
6. If the blanket exterior is conductive (paint, ITO, fog), make sure that it is grounded. Verify with an ohmmeter.

3.2.5.2 Thermal Control Louvers. Bond/ground the thermal control louver blades and axles. The easiest way to bond the blades to the chassis is to have the bimetallic spring bonded electrically at both the blade/axle and the spacecraft structure. Alternatively, place a thin wiper wire from the spacecraft chassis to the axle.

3.2.5.3 Antenna Grounding. Antenna elements should usually be electrically grounded to the structure. Implementation of antenna grounding will require careful consideration in the initial design phase. All metal surfaces, booms, covers, and feeds should be grounded to the structure by wires and metallic screws (dc short design). All waveguide elements should be bonded together electrically with spot-welded connectors and grounded to the spacecraft structure. These elements must be grounded to the Faraday cage at their entry points. Conductive epoxy can be used where necessary, but dc resistance of about 1 Ω should be verified by measurements.

3.2.5.4 Antenna Apertures. Spacecraft radio-frequency (RF) antenna aperture covers should usually be ESD conductive and grounded. Charging and arcing of dielectric antenna dish surfaces and radomes can be prevented by covering them with grounded ESD-conductive material. Antenna performance should be verified with the ESD covering installed.

For a dielectric radome, there have been problems of damage to nearby electronics. Sometimes the radome may be spaced very near low-noise amplifiers (LNAs). If the radome surface charges, electrostatic attraction may draw its surface near the LNAs, and a spark could destroy them; this is a suspected culprit for some on-orbit failures. In such a case, the radome must be spaced far enough away that it cannot damage any LNA or similar nearby electronic devices.

A similar problem exists if there are metal antenna elements in a dielectric matrix, all exposed on the surface. An ESD arc from the dielectric to the antenna element, carried down a coaxial cable to the receiver front end (or transmitter output), can do the same sort of damage. Situations such as this (ESD events caused by surface metals near dielectrics that are carried down to delicate electronics) must be handled with care; filtering or diode protection must be applied to protect the electronics from damage.

Coverings on antenna feeds and parabolas should be considered. Isolated dielectric materials on an antenna system, especially near feed lines, can store excess charge or energy. For example, if there is an isolated dielectric mounted on top of a fiberglass separator that is adjacent to the feed electrical path, there can be discharges directly into the receiver. These dielectrics are special problems because they are on the outside of the spacecraft and have less shielding. Assess each of the region's hazards, and compare to the receiver or LNA ESD sensitivity.

3.2.5.5 Antenna Reflector Surfaces Visible to Space. Grounded, conductive spacecraft charge-control materials should be used on antenna reflector rear surfaces visible to space. Appropriate surface covering techniques must be selected. Such methods include conductive meshes bonded to dielectric materials, silica cloth, conductive paints, or nonconductive (but charge bleeding) paints overlapping grounded conductors. Properly constructed thermal blankets may also accomplish this need to prevent surface charging. Ungrounded array elements, such as for special antenna surfaces, may include ungrounded conductors as a necessary part of their design (e.g., tuned reflector array elements). These may be left ungrounded if analysis shows that the stored energy available from space charging will not affect any possible victims.

3.2.5.6 Transmitters and Receivers. Spacecraft transmitters and receivers should be immune to transients produced by ESDs, including those from dielectrics in the antenna (surface charging) and feed system (internal charging). Transmitter and receiver electrical design must be compatible with the results of spacecraft charging effects. The EMI environment produced by spacecraft ESD should be addressed early in the design phase to permit effective electrical design for immunity to this environment. The transmitter, receiver, and antenna system should be tested for immunity to ESDs near the antenna feed. Consider the possibility of an arc from a dielectric that sparks to the center conductor of a coaxial cable to a delicate receiver or transmitter device at the other end of that coaxial cable. Change the design if necessary. Verification tests should be established by an experienced ESD engineer.

3.2.5.7 Attitude Control Packages. Attitude control electronics packages should be made insensitive to ESD transients. Attitude control systems often require sensors that are remote from electronics packages for Faraday cage shielding. This presents the risk that ESD transients will be picked up and conducted into electronics, especially via the cabling if shielded inadequately. Particular care must be taken to ensure immunity of interface circuits to ESD upset in such cases.

3.2.5.8 Deployed Packages. Deployed packages should be grounded by using a flat ground strap extending the length of the boom to the vehicle structure. Several spacecraft designs incorporate dielectric booms to deploy payloads. The payload electrical system may still require a common ground reference, or the experiment may require a link to some electric potential reference. In these cases, it is recommended that a flat ground strap be used to carry this ground tie to the vehicle structure. Electrical wiring extending from the deployed payload to the spacecraft interior must be carried inside or along the dielectric booms. This wiring should be shielded and the shield grounded at the package end and at the Faraday cage entrance.

3.2.5.9 Ungrounded Materials. Specific items that cannot be grounded because of system requirements should undergo analysis to assure specified performance in the charging environment. Certain space vehicles may contain specific items or materials that must not be grounded. For example, a particular experiment may have a metallic grid or conducting plate that must be left ungrounded. If small, these items may present no unusual spacecraft charging problems; however, this should be verified through analysis.

3.2.5.10 Honeycomb Structures. Honeycomb structures need special grounding methods. Be aware that the aluminum honeycomb interior may be isolated from conductive and grounded face sheets by felt pre-preg adhesive-impregnated material. A small ground wire running across the aluminum honeycomb and pressed against the edge can provide a ground. The conductive face sheets may lose their grounding when they are butted against each other. Develop processes that assure that all metal parts of the honeycomb structure and face sheets will be grounded. After assembly, the inner parts cannot be checked to see if they are grounded.

3.2.5.11 Deliberate or Known Surface Potentials. If a surface on the spacecraft must be charged (e.g., detectors on a science instrument), it should be recessed or shielded so that the perturbation in the surface electrostatic potential is less than 10 V. Scientific instruments that have exposed surface voltages for measurement purposes, such as Faraday cups, require special attention to ensure that the electrostatic fields they create will not disrupt adjacent surface potentials or cause discharges by their operation. They can be recessed so that their fields at the spacecraft surface are minimal or shielded with grounded grids. These

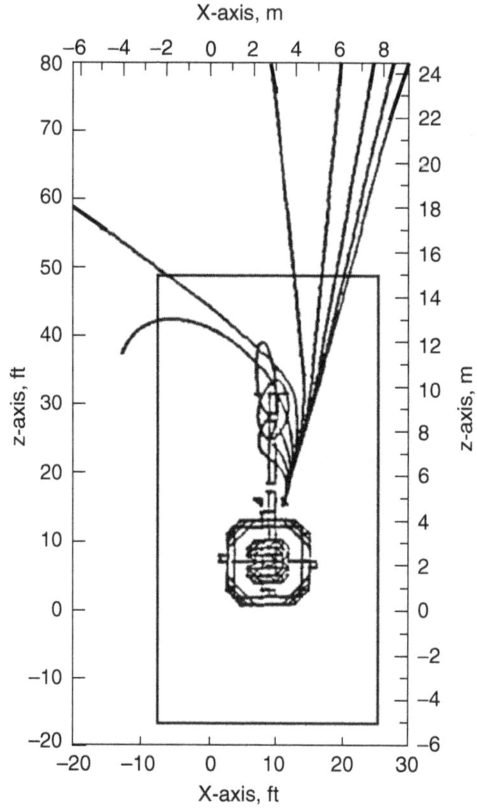

Fig. 3-10. Electron trajectories for *Galileo*. (From [25].)

detector apertures should have a conductive grounded surface around them and in their field of view. An analysis may be necessary to ensure that their presence is acceptable from a charging standpoint and that surrounding surfaces do not affect the measurements.

Figure 3-10 presents an analytic result showing the disturbances in electron paths in the presence of electric fields from spacecraft surface charging, in this case from dielectric surfaces charged by space plasma. Figure 3-10 shows a calculation of particle trajectories distorted by electric fields on parts of the *Galileo* spacecraft. The 10 curves represent paths of 1- to 50-eV electrons, with lines at logarithmically equally spaced energies. The distorted paths of the lower-energy electrons show clearly in this simulation. The design was changed to permit undistorted science measurements.

3.2.5.12 Spacecraft-Generated Plasma Environment. The total plasma environment includes plasma generated by spacecraft electric propulsion (i.e., arc jets, Hall thrusters, and ion thrusters) and possibly other sources. This was shown to be a critical consideration because when thrusters are fired they can surround GEO spacecraft with LEO-type plasma. That plasma can have a major impact on, as a minimum, GEO solar array designs. It is especially important if thrusters are fired during the time a spacecraft is negatively charged by GEO plasma. It can result in unexpected synergistic effects that can lead to ESD events and damage of solar arrays [26].

REFERENCES

[1] A. C. Whittlesey, *Electrical Grounding Architecture for Unmanned Spacecraft*, NASA-HDBK-4001, National Aeronautics and Space Administration, 23 pages, February 17, 1998.

[2] J. Brunson and J. R. Dennison, "Dependence of Resistivity in Low-Density Polyethylene on Space Environment Parameters", *10th Spacecraft Charging Technology Conference*, Biarritz, France, June 18–21, 2007.

[3] C. F. Hoeber, E. A. Robertson, I. Katz, V. A. Davis, and D. B. Snyder, "Solar Array Augmented Electrostatic Discharge in GEO", *17th International Communications Satellite Systems Conference and Exhibit*, Yokohama, Japan, AIAA 98–1401, February 23–27, 1998. This earlier reference is very desirable because of the clarity of exposition (several of its graphics are used in this document), the wide variety of parameters studied, and the numerous physics derivations used in their attempts to explain both the on-orbit experiences and the ground test results, and specific design guidelines to mitigate problems in the future. Later researchers have improved on the contents of this paper in terms of additional test complexity, but it is very readable. The wise person will read later papers and get testing done by persons with recent test experience and knowledge.

[4] R. S. Bever and J. Staskus, "Tank Testing of a 2500-cm^2 Solar Panel," *Spacecraft Charging Technology 1980*, Colorado Springs, Colorado, AFGL-TR-0270/NASA CP-2182, National Aeronautics and Space Administration, pp. 211–227, 1981. This paper supports cerium-doped solar array cover slide charging being lower than fused silica.

[5] C. K. Purvis, H. B. Garrett, A. C. Whittlesey, and N. J. Stevens, *Design Guidelines for Assessing and Controlling Spacecraft Charging Effects*, NASA Technical Paper 2361, National Aeronautics and Space Administration, September 1984. This document has been widely used by practitioners of this art (usually, EMC engineers or radiation survivability engineers) since its publication in 1984. Its contents are limited to surface charging effects. The contents are valid to this day for that purpose. NASA TP-2361 contents have been incorporated into this and NASA-STD-4002, Revision A, with heavy editing. Many of the original details, especially time-variant and multiple-case versions of suggested environments, have been simplified into single worst-case environments in NASA-HDBK-4002, Revision A. Some background material has not been transferred into this document, so the original may still be of interest.

[6] M. Bodeau, "Going Beyond Anomalies to Engineering Corrective Action, New IESD Guidelines Derived from a Root-Cause Investigation", *2005 Space Environmental Effects Working Group Workshop*, Aerospace Corporation, El Segundo, California, 2005. See also Balcewicz et al. [10].

[7] M. Bodeau, "High Energy Electron Climatology That Supports Deep Charging Risk Assessment in GEO", *48th AIAA Aerospace Sciences Meeting*, Orlando, Florida, AIAA 2010-1608, 2010. A fine work with good concepts, explained and illustrated with actual space data and estimates of fluence accumulation versus material resistivity. The author challenges the 0.1-pA/cm^2 and 10-h flux integration guidelines.

[8] Anonymous, *Test Method Standard for Microcircuits*, MIL-STD-883G, U.S. Department of Defense, 716 pages, February 28, 2006.

[9] *Spacecraft Charging*, ECSS-E-ST-20-06C, European Cooperation for Space Standardization, Noordwijk, The Netherlands, 120 pages, July 31, 2008.

[10] P. Balcewicz, J. M. Bodeau, M. A. Frey, P. L. Leung, and E. J. Mikkelson, "Environmental On-Orbit Anomaly Correlation Efforts at Hughes", *Proceedings of the 6th Spacecraft Charging Conference*, Hanscom Air Force Base, Massachusetts, pp. 227–230, November 2–6, 1998.

[11] W. Khachen, J. Suthar, A. Stokes, R. Dollinger, and W. G. Dunbar, "Aerospace-specific Design Guidelines for Electrical Insulation", *IEEE Transactions on Electrical Insulation*, vol. 28, no. 5, pp. 876–886, October 1993. Includes approximate dielectric strengths (electric field stress) for aerospace dielectrics. Dunbar's valuable experience is expressly mentioned in this paper.

[12] W. G. Dunbar, *Design Guide: Designing and Building High Voltage Power Supplies*, Vol. II, AFWAL-TL-88-4143, Wright Patterson Air Base, Ohio, August 1988. This 333-page classic should be on everyone's bookshelf. Dielectric strength considerations are discussed at length.

[13] P. L. Leung and E. Mikkelson, "Mitigation of the Internal Charging Threat Posed by Energetic Electrons Using an Electrically Leaky Coating," *10th Spacecraft Charging Technology Conference*, Biarritz, France, June 18–21, 2007. This is a very nice paper describing the "before" large electrostatic transients accumulated on traditional conformal coating and the far smaller ESD transients with use of their new coating. Leung claims thorough space qualification for this product, but it is company proprietary and not for sale (in response to an audience question).

[14] D. C. Ferguson, *Low Earth Orbit Spacecraft Charging Design Standard*, NASA-STD-4005, National Aeronautics and Space Administration, 16 pages, June 3, 2007.

[15] D. C. Ferguson, *Low Earth Orbit Spacecraft Charging Design Handbook*, NASA-HDBK-4006, National Aeronautics and Space Administration, 63 pages, June 3, 2007.

[16] D. C. Ferguson and G. B. Hillard, "New NASA SEE LEO Spacecraft Charging Design Guidelines: How to Survive in LEO Rather than GEO," *8th Spacecraft Charging Technology Conference*, NASA Marshall Space Flight Center, Huntsville, Alabama, 2003.

[17] I. Katz, V. A. Davis, and D. Snyder, "Mechanism for Spacecraft Charging Initiated Destruction of Solar Arrays in GEO", *36th AIAA Aerospace Sciences Meeting and Exhibit*, AIAA-1998-1002, 1998. This is a seminal text in this area. Be sure to read this document from one of the key experts in the field.

[18] M. J. Mandell, V. A. Davis, B. M. Gardner, and G. Joneward, "Electron Collection by International Space Station Solar Arrays", *8th Spacecraft Charging Conference*, Huntsville, Alabama, 2003.

[19] D. Payan, J.-F. Rousse, E. Daly, D. Ferguson, and S. Lai, Eds., *Proceedings of the 10th Spacecraft Charging Technology Conference*, Biarritz, France (sponsors: ONERA, ESA, NASA, and CNES), 2007.

[20] V. A. Davis, M. J. Mandell, D. L. Cooke, and C. L. Enloe, "High-Voltage Interactions in Plasma Wakes: Simulation and Flight Measurements from the Charge Hazards and Wake Studies (CHAWS) Experiment", *Journal of Geophysical Research*, vol. 104, no. A6, pp. 12,445–12,459, June 1, 1999.

[21] K. Bogus, C. Claassens, and H. Lechte, "Investigations and Conclusions on the ECS Solar Array In-Orbit Power Anomalies", *Proceedings of the 18th IEEE Photovoltaic Specialists Conference*, Las Vegas, Nevada, pp. 368–375, October 21–25, 1985. MARECS-A and ECS-1 experienced partial loss of power after five years on station. In-orbit tests identified the failure to be a short to the panel structure of several sections of the array. This is one of the earliest known references to a flight failure analysis. The analysis resulted in a recommendation to resistively isolate the solar array structure from the spacecraft structure.

[22] V. Inguimbert, L. Levy, F. Bouay, D. Sarrail, G. Migliorero, and P. A. Mausli, "Study of Arc Propagation in Solar Array Drive Mechanisms", *10th Spacecraft Charging Technology Conference*, Biarritz, France, June 18–21, 2007. The authors note parameters of track spacing, housing insulation, direct line of sight between conductors, or inter-track insulation barrier height increase, and brush insulation as design parameters to consider.

[23] H. Brandhorst, J. Rodiek, D. Ferguson, and M. O'Neill, "Stretched Lens Array (SLA): A Proven and Affordable Solution to Spacecraft Charging in GEO", *10th Spacecraft Charging Technology Conference*, Biarritz, France, June 18–21, 2007. This new concept design, based on the DS-1 spacecraft, eliminates many space charging problems that are present on normal solar arrays. It uses cylindrical Fresnel lenses, focusing light on 1-cm-wide triple junction cells (overall efficiency 27%), and over 1000-V differential voltage standoff adjacent cells. It is fully insulated and micrometeoroid-immune. Brandhorst also says that conductive cover glass ($10^9\ \Omega \cdot \text{cm}$) is helpful; may be grounded or may be at the cell potential on an indium tin oxide (ITO) surface.

[24] E. Amorim, D. Payan, R. Reulet, and D. Sarrail, "Electrostatic Discharges on a 1m^2 Solar Array Coupon: Influence of the Energy Stored on Coverglass on Flashover Current", *9th Spacecraft Charging Technology Conference*, Tsukuba, Japan, April 2005. Plasma propagation speed: $0.7–1.1 \times 10^4$ m/s.

[25] M. Harel, *Galileo NASCAP Analysis Report*, D-69473, IOM-5137-82-62 (internal report), Jet Propulsion Laboratory, Pasadena, California, May 1982.

[26] J. J. Likar, A. L. Bogorad, T. R. Malko, N. E. Goodzeit, J. T. Galofaro, and M. J. Mandell, "Interaction of Charged Spacecraft with Electric Propulsion Plume: On Orbit Data and Ground Test Results", *IEEE Transactions on Nuclear Science*, vol. 53, no. 6, pp. 3602–3606, 2006.

[27] D. C. Ferguson, "Plasma Effects on Spacecraft Then and Now: A Welcome to Participants", *Proceedings of the 6th Spacecraft Charging Conference*, Hanscom Air Force Base, Massachusetts, pp. 1–5, November 2–6, 1998.

4 Spacecraft Test Techniques

Spacecraft and systems should be subjected to transient upset tests to verify immunity. It is the philosophy in this document that testing is an essential ingredient in a sound spacecraft charging protection program. In this chapter, the philosophy and methods of testing spacecraft and spacecraft systems are reviewed. The content here is largely unchanged from the analogous section in NASA TP-2361 [1].

4.1 TEST PHILOSOPHY

The philosophy of an ESD test is identical to that of other environmental qualification tests:

1. Subject the spacecraft to an environment representative of that expected.
2. Make the environment applied to the spacecraft more severe than expected as a safety margin to give confidence that the flight spacecraft will survive the real environment.
3. Have a design qualification test sequence that is extensive and includes the following:
   a. Test of all units of hardware
   b. Use of long test durations
   c. Incorporation of as many equipment operating modes as possible
   d. Application of the environment to all surfaces of the test unit
4. Have a flight hardware test sequence of more modest scope, such as deleting some units from test if qualification tests show great design margins; use shorter test durations; use only key equipment operating modes; and apply the environment to a limited number of surfaces.

Ideally, both prototype and flight spacecraft should be tested in a charging simulation facility. They should be electrically isolated from ground and bombarded with electron, ion, and EUV radiation levels corresponding to substorm environment conditions. Systems should operate without upset throughout this test. Generally, there is a reluctance to subject flight hardware to this kind of

Guide to Mitigating Spacecraft Charging Effects, First Edition. Henry B. Garrett, Albert C. Whittlesey.

test. One good reason is the possibility of latent damage (i.e., internal physical damage to circuitry that apparently still functions but that has weakened the hardware and may lead to later failure). For that reason, flight hardware is ESD tested less frequently than is developmental hardware. For the same reason, flight hardware might be subjected to lower test amplitudes as a precaution to demonstrate survivability but without margin.

Because of the difficulty of simulating the actual environment (space vacuum and plasma parameters, including species such as ions, electrons, and heavier ions; mean energy; energy spectrum; and direction), spacecraft charging tests usually take the form of assessing unit immunity to electrical discharge transients. The appropriate discharge sources are based on separate estimates of discharge parameters. Tests at room ambient temperature using radiated and injected transients are more convenient. These ground tests, however, cannot simulate all the effects of the real environment because the transient source may not be in the same location as the region that may discharge and because a spark in air has a slower risetime than a vacuum arc. The sparking device's location and pulse shape must be analyzed to provide the best possible simulation of coupling to electronic circuits. To account for the difference in risetime, the peak voltage might be increased to simulate the dV/dt (time rate of change of the voltage) parameter of a vacuum arc. Alternatively, the voltage induced during a test could be measured and the in-flight noise extrapolated from the measured data.

There are no simple rules to be followed in determining whether or how much to test. General guidance dictates that an engineering version of the hardware be tested in lieu of the flight hardware and that this testing has margins that are more severe than the environment expected. The trade-offs are common to other environmental testing; the main difference is that the ESD- and IESD-specific threats are more difficult to replicate in practical tests than for other environmental disciplines. The dangers in not testing are that serious problems related to surface or internal charging will go undetected and that these problems will affect the survivability of the spacecraft. The best that can be done in the absence of testing is good design supplemented by analysis. Good IESD and surface charging design techniques are always appropriate no matter what the overt environmental threat is, and should be followed as a necessary precaution in all cases. Proper risk assessment involves a well-planned test, predictions of voltage stress levels at key spacecraft components, verification of these predictions during test, checkout of the spacecraft after test, and collaboration with all project elements to coordinate and assess the risk factors.

4.2 SIMULATION OF PARAMETERS

Because ESD test techniques are not well established, it is important to understand the various parameters that must be simulated, at a minimum, to perform an adequate test. On the basis of their possibility of interference to the spacecraft, the following items should be considered in designing tests:

Table 4-1. Examples of Estimated Space-Generated ESD Spark Parameters

ESD Generator	C^a(nF)	V_b^b(kV)	E^c(mJ)	I_{pk}^d(A)	T_R^e(ns)	T_P^f(ns)
Dielectric plate to conductive substrate	20	1	10	2[g]	3	10
Exposed connector dielectric	0.150	5	1.9	36	10	15
Paint on high-gain antenna	300	1	150	150	5	2400
Conversion coating on metal plate (anodize)	4.5	1	2.25	16	20	285
Paint on optics hood	550	0.360	36	18	5	600

Source: [1], with corrections.

[a]Capacitance computed from surface area, dielectric thickness, and dielectric constant.

[b]Breakdown voltage computed from dielectric thickness and material breakdown strength.

[c]Energy computed from $E = \frac{1}{2}CV^2$.

[d]Peak current estimated based on measured data; extrapolation based on square root of area.

[e]Discharge current risetime measured and deduced from test data.

[f]Discharge current pulse width to balance total charge on capacitor.

[g]Replacement current in longer ground wire; charge is not balanced.

- Spark location
- Radiated fields or structure currents
- Area, thickness, and dielectric strength of the material
- Total charge involved in the event
- Breakdown voltage
- Current waveform [risetime, width, falltime, and rate of rise (in A/s)]
- Voltage waveform [risetime, width, falltime, and rate of rise (in V/s)]

Table 4-1 shows typical values calculated for representative spacecraft. The values listed in this table were compiled from a variety of sources, mostly associated with the *Voyager* and *Galileo* spacecraft. The values for each item (e.g., those for the dielectric plate) have been assembled from the best available information and made into a more or less self-consistent set of numbers. The process is described in the footnotes to Table 4-1. References [2] and [3] contain further descriptions and discussion.

4.3 GENERAL TEST METHODS

4.3.1 ESD-Generating Equipment

Several representative types of test equipment are tabulated in Table 4-2 and are described later. Where possible, typical parameters for that type of test are listed.

Table 4-2. Examples of Several ESD Sources

ESD Generator Test Simulation	C (nF)	V_b (kV)	E (mJ)	I_{pk} (A)	T_R (ns)	T_P (ns)
MIL-STD-1541A (auto coil)[a]	0.035	19	6	80	5	20
Flat plate 20 cm × 20 cm at 5 kV, 0.8 mm (3 mils) Mylar® insulation	14	5	180	80	35	880
Flat plate with lumped-element capacitor	550	0.450	56	15	15	[b]
Capacitor direct injection	1.1	0.32	0.056	1	3–10	20
Capacitor arc discharge	60	1.4	59	1000	[c]	80
Commercial ESD tester	0.15	20	30	130	5	22

Source: [1], with corrections.
[a]Parameters were measured on one unit similar to the MIL-STD-1541A, *Electromagnetic Compatibility Requirements for Space Systems* [4], design.
[b]*RC* time constant decay can be adjusted with an external resistor in the circuit.
[c]Value uncertain.

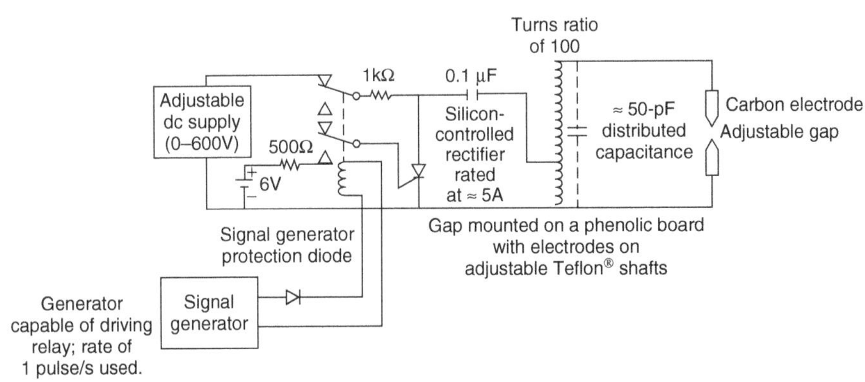

Fig. 4-1. MIL-STD-1541A arc source.

4.3.1.1 MIL-STD-1541A Arc Source. The schematic and usage instructions for the MIL-STD-1541A [4] arc source are presented in Fig. 4-1 and typical gap-spacings and voltage breakdown levels are listed in Table 4-3. The arc source can be manufactured relatively easily and can provide the parameters necessary to simulate a space-caused ESD event. The only adjustable parameter for the MIL-STD-1541A arc source, however, is the discharge voltage achieved by adjusting the discharge gap and, if necessary, the adjustable dc supply to the discharge

Table 4-3. Typical Gap Spacing and Voltage Breakdown Levels

Gap (mm)	V_b (kV)	Approximate Energy Dissipated (μJ)
1	1.5	56.5
2.5	3.5	305
5	6	900
7.5	9	2000

capacitor. As a result, peak current and energy vary with the discharge voltages. Since the risetime, pulse width, and falltime are more or less constant, the voltage and current rates of rise and fall are not independent parameters. This permits some degree of flexibility in planning tests but not enough to cover all circumstances. Recent versions of MIL-STD-1541 no longer refer to this test method.

4.3.1.2 Flat-Plate Capacitor. A flat-plate capacitor made of aluminum foil over an insulator can be used in several circumstances. Examples of spacecraft areas that can be simulated by a flat-plate capacitor are thermal blanket areas, dielectric areas such as calibration targets, and dielectric areas such as nonconductive paints. The chief value of a flat-plate capacitor is to permit a widespread discharge to simulate the physical path of current flow. This can be of significance where cabling or circuitry is near the area in question. Also, the larger size of the capacitor plates allows them to act as an antenna during discharge, producing significant radiated fields.

Table 4-2 shows one example of the use of a flat-plate capacitor. Several parameters can be varied, chiefly the area and the dielectric thickness; both of these affect the capacitance, the discharge current, and the energy. The discharge voltage of the flat plate can be controlled by using a needle-point discharge gap at its edge that is calibrated to break down before the dielectric. This gap also affects discharge energy. In this manner, several mechanical parameters can be designed to yield discharge parameters more closely tailored to those expected in space.

The difficulties of this method include the following:

1. The test capacitor is usually not as close to the interior cabling as the area that it is intended to simulate (e.g., it cannot be placed as close as the paint thickness).
2. The capacitance of the test capacitor may be less than that of the area it is intended to simulate. To avoid uncontrolled dielectric breakdown in the test capacitor, its dielectric may have to be thicker than the region it simulates. If so, the capacitance will be reduced. The area of the test capacitance can be increased to compensate, but then the size and shape will be less realistic.

4.3.1.3 Lumped-Element Capacitors. Use of lumped-element capacitors (off-the-shelf manufactured capacitors) can overcome some of the objections raised about flat-plate capacitors. They can have large capacitances in smaller areas and thus supplement a flat-plate capacitor if it alone is not adequate. The deficiencies of lumped-element capacitors are as follows:

1. They generally do not have the higher breakdown voltages (greater than 5 kV) needed for ESD tests.
2. Some have a high internal resistance and cannot provide the fast risetimes and peak currents needed to simulate ESD events.

Generally, the lumped-element capacitor discharge would be used most often in lower-voltage applications to simulate painted or anodized surface breakdown voltages and in conjunction with the flat-plate capacitors.

4.3.1.4 Other Source Equipment. Wilkenfeld et al. [5] describe several other similar types of ESD simulators. The review is a useful document if further descriptions of ESD testing are desired.

4.3.1.5 Switches. A wide variety of switches can be used to initiate the arc discharge. At low voltages, semiconductor switches can be used. The MIL-STD-1541A arc source uses a silicon-controlled rectifier to initiate the spark activity on the primary of a step-up transformer; the high voltage occurs at an air spark gap on the transformer's secondary. Also at low voltages, mechanical switches can be used (e.g., to discharge modest-voltage capacitors). The problem with mechanical switches is their bounce in the early milliseconds. Mercury-wetted switches can alleviate this problem to a degree.

For high-voltage switching in air, a gap made of two pointed electrodes can be used as the discharge switch. Place the tips pointing toward each other and adjust the distance between them to about 1 mm/kV of discharge voltage. The gap must be tested and adjusted before the test, and it must be verified that breakdown occurred at the desired voltage. For tests that involve varying the amplitude, a safety gap connected in parallel is suggested. The second gap should be set securely at the maximum permissible test voltage. The primary gap can be adjusted during the test from zero to the maximum voltage desired without fear of inadvertent overtesting. Do the test by charging the capacitor (or triggering the spark coil) and relying on the spark gap to discharge at the proper voltage.

The arc source's power supply must be isolated sufficiently from the discharge so that the discharge is a transient and not a continuing arc discharge. A convenient test rate is once per second. To accomplish this rate, it is convenient to choose the capacitor and isolation resistor's resistance–capacitance time constant to be about 0.5 s and to make the high-voltage power supply output somewhat higher than the desired discharge voltage. For tests that involve a fixed discharge voltage, gas discharge tubes are available with fixed breakdown voltages. The advantage of the gas discharge tube over needle points in air is its faster risetime and its very repeatable discharge voltage. The gas discharge tube's dimensions (5 to 7 cm or longer) can cause more RF radiation than a smaller set of needle-point air gaps.

Another type of gas discharge tube is the triggered gas discharge tube. This tube can be triggered electronically, much as the gate turns on a silicon-controlled rectifier. This method has the added complexity of the trigger circuitry. Additionally, the trigger circuitry must be properly isolated so that discharge currents are not diverted by the trigger circuits.

4.3.2 Methods of ESD Applications

The ESD energy can range from very small to large (as much as 1 J but usually millijoules). The methods of application can range from indirect (radiated) to direct (applying the spark directly to a piece part). In general, the method of application should simulate the expected ESD source as much as possible. Several typical methods are described here.

*4.3.2.1 **Radiated Field Tests.*** The sparking device can be operated in air at some distance from the component. This technique can be used to check for RF interference to communications or surveillance receivers as coupled into their antennas. It can also be used to check the susceptibility of scientific instruments that may be measuring plasma or natural radio waves. Typical RF-radiated spectra are shown in Fig. 4-2.

*4.3.2.2 **Single-Point Discharge Tests.*** Discharging an arc onto the spacecraft surface or a temporary protective metallic fitting with the arc current return wire in close proximity can represent the discharge and local flowing of arc currents. This test is more severe than the radiated test, since it is performed immediately

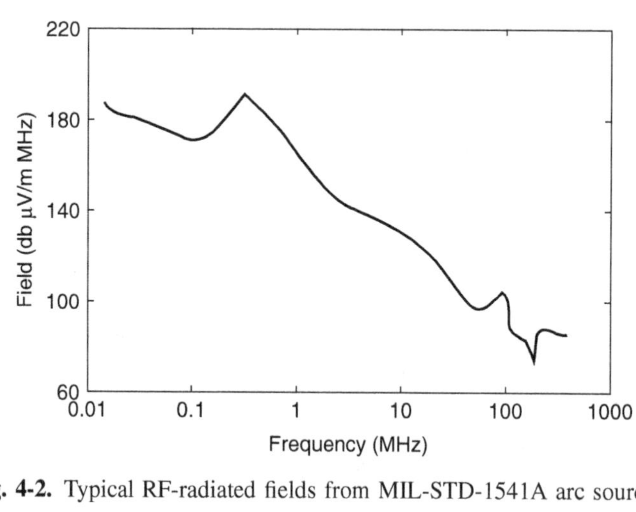

Fig. 4-2. Typical RF-radiated fields from MIL-STD-1541A arc sources.

adjacent to the spacecraft rather than some distance away. This test simulates only local discharge currents; it does not simulate blow-off of charges which cause currents in the entire structure of the spacecraft.

4.3.2.3 ***Structure Current Tests.*** The objective of structure current testing is to simulate blow-off of charges from a spacecraft surface. If a surface charges and an ESD results, the spark may vaporize and remove material and charges mechanically without local charge equalization. In such a case, the remaining charge on the spacecraft will redistribute itself and cause structural currents.

Defining the actual blow-off currents and the paths they take is difficult. Nevertheless, it is appropriate to do a structure current test to determine the spacecraft susceptibility using test currents and test locations supported by analysis as illustrated in Section 4.2 and Table 4-1. Typically, such a test would be accomplished by using one or more of the following current paths (Fig. 4-3):

a. Diametrically opposed locations (through the spacecraft)
b. Protuberances (from landing foot to top, from antenna to body, and from thruster jets to opposite side of body)
c. Extensions or booms (from end of sensor boom to spacecraft chassis and from end of solar panel to spacecraft chassis)
d. From launch attachment point to other side of spacecraft

The tests using current paths a and d are of a general nature while those using current paths b and c simulate probable arc locations on at least one end of the current path. These test points include thrusters, whose operation can trigger an incipient discharge, and also landing feet and the attachment points, especially if used in a docking maneuver, when they could initiate a spark to the mating spacecraft. Test c is an especially useful test. Solar panels often have glass (nonconductive) cover slides, and sensors may have optics (nonconductive) that can cause an arc discharge. In both cases, any blow-off charge would be replaced by a current in the supporting boom structure that could couple into cabling in

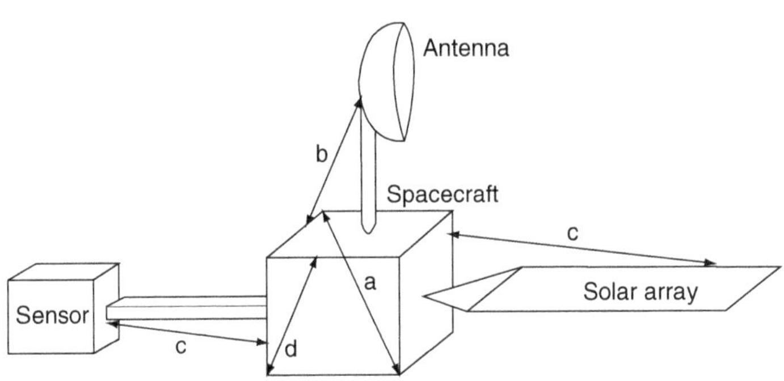

Fig. 4-3. Paths for ESD currents through a structure.

the boom. This phenomenon is possibly the worst-case event that could occur on the spacecraft because the common length of the signal or power cable near the arc current is the longest on the spacecraft.

4.3.2.4 Unit Testing.

4.3.2.4.1 General. Unit ESD testing serves the same purpose as it serves in standard environmental testing (i.e., it identifies design deficiencies at a stage when the design is more easily changed). It is, however, very difficult to provide a realistic determination of the unit's environment as caused by an ESD on the spacecraft.

A unit testing program could specify a single ESD test for all units or could provide several general categories of test requirements. The following test categories are provided as a guide:

1. Internal units (general) must survive, without damage or disruption, the MIL-STD-1541A arc source test (discharges to the unit but no arc currents through the unit's chassis).
2. External units mounted outside the Faraday cage (usually, exterior sensors) must survive the MIL-STD-1541A arc source at a 5-kV level with discharge currents passing from one corner to the diagonally opposite corner (four pairs of locations).
3. For units near a known ESD source (e.g., solar cell cover slides and Kapton® thermal blankets), the spark voltage and other parameters must be tailored to be similar to the spark expected from that dielectric surface. For solar cells it is important that the arc injection point be at the edge of a cell rather than at an interconnect or bypass diode. This is because solar cell damage happens because of high current densities at cell edges rather than because of currents flowing through the cell.

4.3.2.4.2 Unit Test Configuration. ESD tests of the unit (subsystem) can be performed with the subsystem configured as it would be for a standard EMC-radiated susceptibility test. The unit is placed on and electrically bonded to a grounded copper-topped bench. The unit is cabled to its support equipment, which is in an adjacent room. The unit and cabling should be of flight construction with all shields, access ports, and so on, in flight condition. All spare cables should be removed.

4.3.2.4.3 Unit Test Operating Modes. The unit should be operated in all modes appropriate to the ESD arcing situation. Additionally, the unit should be placed in its most sensitive operating condition (amplifiers in highest gain state, receivers with a very weak input signal) so that the likelihood of observing interference from the spark is maximized. The unit should also be exercised through its operating modes to assure that mode change commands are possible in the presence of arcing.

4.3.2.5 Spacecraft Testing. The system-level test will provide the most reliable determination of the performance expected from a space vehicle in the charging environment. Such a test should be conducted on a representative spacecraft before exposing the flight spacecraft to ensure that there will be no inadvertent overstressing of flight units. A detailed test plan must be developed that defines test procedures, instrumentation, test levels, and parameters to be investigated. Test techniques will probably involve current flow in the spacecraft structure. Tests can be conducted in ambient environments, but screen rooms with electromagnetic dampers are recommended. MIL-STD-1541A system test requirements and radiated EMI testing are considered to be a minimal sequence of tests.

The spacecraft should be isolated from ground. Instrumentation must be screened electrically from the discharge test environment and must be carefully chosen so that instrument response is not confused with spacecraft response. The spacecraft and instrumentation should be on battery power. Complete spacecraft telemetry should be monitored. Voltage probes, current probes, E- and H-field current monitors, and other sensors should be installed at critical locations. Sensor data should be transmitted with fiber-optic data links for best results. Oscilloscopes and other monitoring instruments should be capable of resolving the fast response to the discharges that is expected (usually, less than 250 MHz of frequency).

The test levels should be determined from an analysis of discharging behavior in the substorm environment. It is recommended that full-level testing, with test margins, be applied to structural, engineering, or qualification models of spacecraft with only reduced levels applied to flight units. The test measurements (e.g., structural currents, harness transients, and upsets) are the key system responses that are to be used to validate predicted behavior.

4.3.2.5.1 General. Spacecraft testing is generally performed in the same fashion as unit testing. A test plan of the following sort is typical (see Fig. 4-3):

1. The MIL-STD-1541A radiated test is applied around the entire spacecraft.
2. Spark currents from the MIL-STD-1541A arc source are applied through spacecraft structure from launch vehicle attachment points to diagonally opposite corners.
3. ESD currents are passed down the length of booms with cabling routed along them (e.g., sensor booms or power booms). Noise pickup into cabling and circuit disruption are monitored.
4. Special tests are devised for special situations. For example, dielectric regions such as quartz second-surface mirrors, Kapton® thermal blankets, and optical viewing windows should have ESD tests applied on the basis of their predicted ESD characteristics.

Examples of system-level ESD current injection test results are shown in Fig. 4-4. The MIL-STD-1541A ESD waveform generator was measured directly

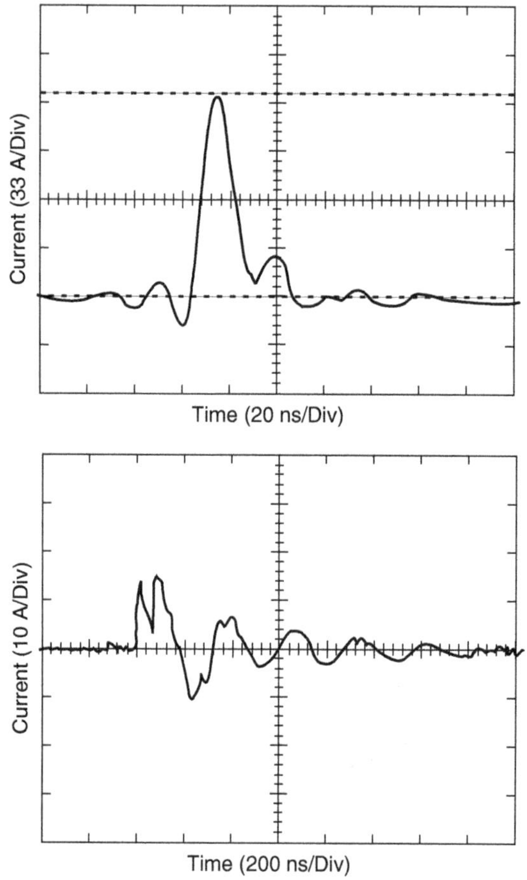

Fig. 4-4. Examples of system-level ESD test waveforms [units of time and current are noted per division ("Div") tic mark].

with very short leads on the output. The peak current is about 66 A and the risetime about 5.2 ns, with a time base of 20 ns per division (ns/div). The waveform was measured during a system-level test. The current was applied via 9 m of attachment wiring (two 4.5-m lengths) from the same MIL-STD-1541A sparker to the top of a spacecraft, with the current return at the solar array drive on the body of the spacecraft. Because of inductance in the long leads, the risetime has increased to 40 ns, the peak current is now 15 A, and the time base is 200 ns/div. (Scale factors in these historic pictures are different in each picture and include attenuations and probe factors.)

4.3.2.5.2 Spacecraft Test Configuration. The spacecraft ESD testing configuration ideally simulates a 100% flight-like condition. This may be difficult because of the following considerations:

- Desire for ESD diagnostics in the spacecraft
- Nonrealistic power system (no solar array)
- Local rules about grounding the spacecraft to facility ground
- Cost and schedules to completely assemble the spacecraft for the test and later disassemble it if failures or anomalies occur
- The possible large capacitance to ground of the spacecraft in its test fixture
- ESD coupling onto nonflight test cabling
- A fear of immediate or latent damage to the spacecraft

4.3.2.5.3 Test Diagnostics. To obtain more information about circuit response than can be obtained by telemetry, it is common to use an oscilloscope to measure induced voltages related to the ESD test sparks at key circuits. If implemented improperly, the very wires that access the circuits and exit the spacecraft to test equipment (e.g., oscilloscopes) will act as antennas and show noise that never would be present without those wires.

Two approaches have been used with some success. The first is using conventional oscilloscope probes with great care. Long oscilloscope probes (3 m) were procured from Tektronix. For the circuits being monitored, a small tee breakout connector was fabricated and inserted at the connector nearest the circuit. Two oscilloscope probes were attached to each circuit's active and return wires, and the probe tips were grounded to satellite structure in the immediate vicinity of the breakout tee. The probe grounds were less than 15 cm from the probe tip. The signal was measured on a differential input of the oscilloscope. Before installation, the probes were capacitively compensated to their respective oscilloscope preamplifiers, and it was verified that their common-mode voltage rejection was adequate (normal good practice). The two probe leads were twisted together and routed along metal structure inside the satellite until they could be routed out of the main chassis enclosure. They were then routed (still under thermal blankets) along the structure to a location as remote as possible from any ESD test location and, finally, routed to the oscilloscope. The oscilloscopes were isolated from building ground by isolation transformers. Clearly, this method permits monitoring only a few circuits [3].

A second method of monitoring ESD-induced voltage waveforms on internal circuits is the use of battery-powered devices that convert voltages to light-emitting diode (LED) signals. The LED signals can be transmitted by fiber optics to exterior receiving devices, where the voltage waveform is reconstructed. As with the oscilloscope probes, the monitoring device must be attached to the wires carefully with minimal disturbance to circuit wiring. The fiber-optic cable must be routed out of the satellite with minimal disturbance. The deficiency of such a monitoring scheme is that the sending device must be battery powered, turned on, and installed in the spacecraft before spacecraft buildup; and it must operate for the duration of the test. The need for batteries and the relatively high-power consumption of LED interface circuits severely restrict this method.

Another proposed way to obtain circuit response information is to place peak-hold circuitry (tattle-tales) at key circuit locations, installed as described above. This method is not very useful because the only datum presented is that a certain peak voltage occurred. There is no evidence that the ESD test caused it, and there is no way to correlate that voltage with any one of the test sequences. For analysis purposes, such information is worthless.

4.3.2.5.4 Use of External (Nonflight) Power Supplies. Spacecraft using solar cells or nuclear power supplies often must use support equipment power supplies for ground test activities and thus are not totally isolated from ground. In such cases, the best work-around is to use an isolated and balanced output power supply with its wires routed to the spacecraft at a height above ground to avoid stray capacitance to ground. The power wires should be shielded to avoid picking up stray radiated ESD noise; the shields should be grounded at the support equipment end of the cable only.

4.3.2.5.5 Facility Grounding. To simulate flight, the spacecraft should be isolated from ground. Normal test practice dictates an excellent connection to facility ground. For the purpose of the ESD test, a temporary ground of 0.2 to 2 MΩ or more will isolate the spacecraft. Generally, 0.2 to 2 MΩ is sufficient grounding for special test circumstances of limited duration and can be tolerated by the safety or quality assurance organization for the ESD test.

4.3.2.5.6 Cost and Schedules to Assemble and Disassemble Spacecraft. Often, testing is done in the most compact form possible, attempting to interleave several tasks at one time or to perform tasks in parallel. This practice is incompatible with the needs of ESD testing and must be avoided. A thermal-vacuum test, for example, is configured like the ESD test but has numerous (nonflight) thermocouple leads penetrating from the interior to the exterior of the spacecraft. These leads can act as antennas and bring ESD-caused noise into satellite circuitry where it never would have been. Dynamic (shake table) test configurations have the same problem with the accelerometers.

4.3.2.5.7 Spacecraft Capacitance to Ground During Test. If stray capacitance to facility ground is present during the ESD test, it will modify the flow of ESD currents. For a better test, the spacecraft should be physically isolated from facility ground. It can be shown that raising a 1.5-m-diameter spherical satellite 0.5 m off the test flooring reduces the stray capacitance nearly to that of an isolated satellite in free space. A dielectric (e.g., wood) support structure can be fabricated for the ESD test and will provide the necessary capacitive isolation.

4.3.2.5.8 ESD Coupling onto Nonflight Test Cabling. One method of reducing ESD coupling to and from the spacecraft on nonflight test wiring is the use of ferrite beads on all such wiring. The most realistic approach is to have no nonflight cabling, leaving only information that would be visible while in flight, at the expense of extra diagnostic information.

REFERENCES

[1] C. K. Purvis, H. B. Garrett, A. C. Whittlesey, and N. J. Stevens, *Design Guidelines for Assessing and Controlling Spacecraft Charging Effects*, NASA Technical Paper 2361, National Aeronautics and Space Administration, September 1984. This document has been used widely by practitioners of this art (usually, EMC engineers or radiation survivability engineers) since its publication in 1984. Its contents are limited to surface charging effects. The contents are valid to this day for that purpose. NASA TP-2361 contents have been incorporated into this and NASA-STD-4002, Revision A, with heavy editing. Many of the original details, especially time-variant and multiple-case versions of suggested environments, have been simplified into single worst-case environments in NASA-HDBK-4002, Revision A. Some background material has not been transferred into this document, so the original may still be of interest.

[2] P. L. Leung, G. H. Plamp, and P. A. Robinson, Jr., "Galileo Internal Electrostatic Discharge Program," *Spacecraft Environmental Interactions Technology 1983*, Colorado Springs, Colorado, October 4–6, NASA CP-2359/AFGL-TR-85-0018, National Aeronautics and Space Administration, pp. 423–435, 1983. This paper, documented in the 1985 publication and presented at the 4th Spacecraft Charging Technology Conference, describes a very neat and clear test that measures the effect of line lengths and circuit board metal areas in the resulting ESD amplitude. It also measures the amplitude of ESD transients from electron beam charging on 50-Ω loads from various conductors.

[3] A. C. Whittlesey, "Voyager Electrostatic Discharge Protection Program," *IEEE International Symposium on EMC*, Atlanta, Georgia, pp. 377–383, 1978.

[4] *Electromagnetic Compatibility Requirements for Space Systems*, MIL-STD-1541A (USAF), U.S. Air Force, 42 pages, December 30, 1987.

[5] J. M. Wilkenfeld, B. L. Harlacher, and D. Mathews, *Development of Electrical Test Procedures for Qualification of Spacecraft Against EID*, Vol. II, *Review and Specification of Test Features*, NASA CR-165590, National Aeronautics and Space Administration, April 1982. EID refers to electron-induced discharge, a term used in those days. A very good survey of spacecraft ESD test methodology as of 1982. Includes the methods used for several satellite systems and provides a comparison and evaluation of their relative merits. This is a fundamental document for anyone in the spacecraft ESD test business.

5 Control and Monitoring Techniques

5.1 ACTIVE SPACECRAFT CHARGE CONTROL

Charge control devices are a means of controlling spacecraft potential. Various active charged-particle emitters have been and are being developed and show promise of controlling spacecraft potential in the space plasma environment. At this time, only neutral plasma devices (both ion and electron emitters) have demonstrated the ability to control spacecraft potential in geomagnetic substorms. These devices are sometimes recommended for charge control purposes [1,2]. Plasma contactors are currently the most widely used charge control devices.

Emitted particles constitute an additional term in the current balance of a spacecraft. Because the ambient current densities at geosynchronous altitude are quite small, emitting small currents from a spacecraft can have a strong effect on its potential, as has been demonstrated on ATS-5, ATS-6, SCATHA, and other spacecraft. However, devices that emit particles of only one electric charge (e.g., electrons) are not suitable for active potential control applications unless all spacecraft surfaces are conducting. Activation of such a device will result in a rapid change of spacecraft potential. Differential charging of any insulating surfaces will occur, however, and cause potential barrier formation near the emitter. Emission of low-energy particles can then be suppressed. Higher-energy particles can escape, but their emission could result in the buildup of large differential potentials. Conversely, devices that emit neutral plasmas or neutralized beams (e.g., hollow cathode plasma sources or ion engines) can maintain spacecraft potentials near plasma ground and suppress differential charging. These are therefore a possible type of charge control devices at the cost of reliability and complexity.

5.2 ENVIRONMENTAL AND EVENT MONITORS

The occurrence of environmentally induced discharge effects in spacecraft systems is usually difficult to verify. Often, the only thing known about an anomaly is that it occurred at some spacecraft time. Since most spacecraft are not well

Guide to Mitigating Spacecraft Charging Effects, First Edition. Henry B. Garrett, Albert C. Whittlesey.

instrumented for environmental effects, the state of the environment at the time of the anomaly typically has to be inferred from ground observatory data. These environmental data are not necessarily representative of the environment at the spacecraft location; in fact, the correlation is generally poor.

This problem could be addressed if spacecraft carried a set of environmental monitors (e.g., a simple monitor set designed to measure the characteristic energy and current flux as well as to determine transients on harness positions within the spacecraft) [3]. This would allow a correlation between the onset of the charging environment and possible transients induced on the electronic systems. Representative packages weigh about 1 kg and use 2 to 3 W of power. One commercially available system is the Amptek Compact Environmental Anomaly Sensor (CEASE) package, which measures total radiation dose, radiation dose rate, surface dielectric charging, deep dielectric charging, and single-event effects (http://www.amptek.com/pdf/cease.pdf). Such environmental sensors would be on outside surfaces and preferably in shade. Even more sophisticated packages are available that make detailed scientific measurements of the environment. For example, ion particle detectors in the range 10 to 50 keV are used to sense the onset of geomagnetic substorms. Transient monitors capable of measuring the pulse characteristics have also been used [4]. These systems require larger weight and power budgets, but they do provide better data.

Spacecraft charging effect monitors require data analysis support to produce the desired results. If they were carried on a number of operational satellites, the technology community would be able to obtain a statistical base relating charging to induced transients. The operational people, on the other hand, would be able to tell when charging is of concern, to establish operational procedures to minimize detrimental effects, and to separate system malfunctions from environmentally induced effects. It is recommended that monitor packages be carried on all geosynchronous spacecraft. These packages should consist, at a minimum, of a dosimeter, energetic plasma environment detector, surface potential monitor, and transient voltage pulse detector. Various types of IESD monitors are currently in development and should be seriously considered also.

REFERENCES

[1] C. K. Purvis and R. O. Bartlett, "Active Control of Spacecraft Charging," *Space Systems and Their Interactions with the Earth's Space Environment*, H. B. Garrett and C. P. Pike, Eds., AIAA Press, New York, pp. 299–317, 1980.

[2] R. C. Olsen and E. C. Whipple, *Active Experiments in Modifying Spacecraft Potential: Results from ATS-5 and ATS-6, May 1977–February 1979*, NASA CR-159993, National Aeronautics and Space administration (produced by the University of California, San Diego, La Jolla, California), March 1979.

[3] J. C. Sturman, *Development and Design of Three Monitoring Instruments or Spacecraft Charging*, NASA-TP-1800, National Aeronautics and Space Administration, September 1981.

[4] H. C. Koons, "Aspect Dependence and Frequency Spectrum of Electrical Discharges on the P78-2 (SCATHA) Satellite," *Spacecraft Charging Technology 1980*, NASA CP-2182/AFGL-TR81-0270, National Aeronautics and Space Administration, pp. 478–492, 1981.

6 Material Notes and Tables

This chapter has been included to place material resistivity, density, and dielectric strength properties in one convenient place for ESD analysts. These lists contain spacecraft materials that might often be considered when doing penetrating electron-charging analyses, charge-accumulation analyses, and breakdown estimates. The lists are generally correct, but the reader should recheck the parameters, especially the resistivity and dielectric strength parameters, for any detail work.

6.1 DIELECTRIC MATERIAL LIST

The partial list of basic dielectric material properties in Table 6-1 is provided for illustration and reader convenience only. Data were taken from references [1,2] and other sources, including manufacturer data sheets. Some of the data may only be specified minimum or maximum limits and not typical values; actual resistivity values may differ by many orders of magnitude (e.g., FR4). Note that dielectric strength is always specified as a function of thickness and may be extrapolated to other thicknesses roughly as the inverse square root of the thickness. Each project must be responsible for compiling its own list based on the most current and relevant data. Reference [3] contains lists of dielectric properties for materials not included here. Other often-significant effects not tabulated here include temperature, radiation-induced conductivity, and electric field-induced conductivity.

Figure 6-1 shows how resistivity and the dielectric constant together combine to determine material time constants, indicating relative desirability for ESD-sensitive applications. Suggested break points are "safe" (difficult to accumulate charge) with time constants less than 3 h, "dangerous" (too resistive and likely to cause on-orbit ESD issues in space plasma environments) with time constants greater than 30 h, and the uncertain/marginal region between. The boxes labeled Kapton® and Teflon® illustrate their possible ranges of resistivity. From this chart it can be seen that both are undesirable from an ESD standpoint.

Guide to Mitigating Spacecraft Charging Effects, First Edition. Henry B. Garrett, Albert C. Whittlesey.

Table 6-1. Dielectric Material Characteristics for Internal Charging Studies[a]

Parameter/ Material (units)	Relative Dielectric Constant[b]	Dielectric Strength[c] (V/mil@ mils)	DC Volume Resistivity ($\Omega \cdot$ cm)[d]	Density (g/cm^3)/ Density in Relation to Aluminum	Time Constant[e] (As Noted)
Ceramic (Al_2O_3)	8.8	340@125	$> 10^{12}$	2.2/0.81	> 0.78 s
Delrin	3.5	380@125	10^{15}	1.42/0.52	310 s (5.2 min)
FR4	4.7	420@62	$> 4 \times 10^{14}$	1.78/0.66	> 141 s
Kapton	3.4	7000@1	$\sim 10^{18}$ to 10^{19}	1.4/0.51	3.5 d
	—	580@125	$\sim 10^{18}$ to 10^{19}	1.4/0.51	3.5 d
Mylar®	3	7000@1	10^{18}	1.4/0.51	3.1 d
Polystyrene	2.5	5000@1	10^{16}	1.05/0.39	37 min
Quartz, fused	3.78	410@250	$> 10^{19}$	> 2.6	> 38 d
Teflon® (generic)[f]	2.1	2000–5000 @1	$\sim 10^{18}$ to 10^{19}	2.1/0.78	2.1 d
	—	500@125	$\sim 10^{18}$ to 10^{19}	2.1/0.78	2.1 d

(Blank lines below are for reader's notes and additions.)

[a] If the numbers in the table are "greater than," the actual time constants could be greater than shown (calculated). The numbers here are for room temperature. At low temperatures, the resistivity values may become much greater and the time constants for charge bleed-off can be much greater.
[b] Permittivity (dielectric constant) = relative dielectric constant $\times 8.85 \times 10^{-12}$ F/m.
[c] $\sim$508 V/mil is the same as 2×10^7 V/m.
[d] Resistivity ($\Omega \cdot$ m) = resistivity ($\Omega \cdot$ cm)/100.
[e] Time constant (s) = permittivity (F/m) x resistivity ($\Omega \cdot$ m).
[f] Generic numbers for Teflon®. Polytetrafluoroethylene [(PTFE) (Teflon®)] and fluorinated ethylene propylene (Teflon® FEP) are common forms in use for spacecraft.

6.2 CONDUCTOR MATERIAL LIST

Table 6-2 shows the resistivity and density information for some conductors. References are the same as in Table 6-1 (from mixed sources for illustration only). The partial list of basic conductor characteristics in Table 6-2 is provided for illustration and reader convenience only.

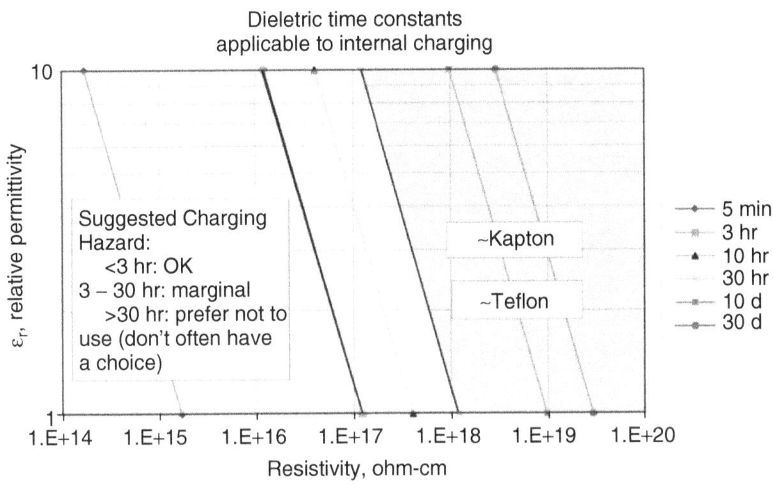

Fig. 6-1. Safe, intermediate, and possibly hazardous dielectric materials based on resistivity and dielectric constant and resulting time constant. (The Kapton® and Teflon® boxes illustrate the uncertainty range for space applications; see the text.) (*See insert for color representation of the figure.*)

Table 6-2. Conductor Characteristics for Charging Studies (Approximate)[a]

Parameter/ Material Units	DC Volume Resistivity[b] (Ω-cm ($\times 10^{-6}$))	DC Volume Resistivity[b] (relative to Al)	Density (g/cm^3)	Density (relative to Al)
Aluminum	2.62	1	2.7	1
Aluminum honeycomb	Variable	Variable	~ 0.049	~ 0.02
Brass (70–30)	3.9	1.49	8.5	3.15
Carbon graphite	5–30	1.9–11.45	1.3–1.95	0.48–0.72
Copper	1.8	0.69	8.9	3.3
Gold	2.44	0.93	19.3	7.15
Graphite–epoxy	Variable	Variable	1.5	0.56
Invar	81	30.9	8.1	3
Iron–steel	9–90	3.43–34.3	7.87	2.91
Lead	98	37.4	11.34	4.2
Kovar A	284	108.4	~7.8	~2.89
Nickel	7.8	2.98	8.9	3.3
Magnesium	4.46	1.7	1.74	0.64
Silver	1.6	0.61	10.5	3.89
Stainless steel	90	34.35	7.7	2.85
Tantalum	13.9	5.3	16.6	6.15
Titanium	48	18.3	4.51	1.67
Tungsten	5.6	2.14	18.8	6.96

(*continued*)

Table 6-2. ***(Continued)***

Parameter/ Material Units	DC Volume Resistivity (Ω-cm (× 10^{-6}))	DC Volume Resistivity (relative to Al)	Density (g/cm^3)	Density (relative to Al)
(Blank lines below are for reader's notes and additions.)				

[a]See the text for references and accuracies. Densities from various sources match well; resistivities may vary.

[b]Resistivity (Ω·m) = resistivity (Ω · cm)/100.

REFERENCES

[1] H. P. Westman, Ed., *Reference Data for Radio Engineers,* 5th Ed.,, Howard Sams & Co., Inc., Indianapolis, Indiana, 1968.

[2] W. T. Shugg, *Handbook of Electrical and Electronic Insulating Materials*, 2nd ed.,, IEEE Press, Piscataway, New Jersey, June 19, 1995. An excellent general reference for dielectric characteristics written by a leader in the field.

[3] A. R. Frederickson, D. B. Cotts, J. A. Wall, and F. L. Bouquet, *Spacecraft Dielectric Material Properties and Spacecraft Charging,* AIAA Progress in Astronautics and Aeronautics, Vol. 107, American Institute of Aeronautics and Astronautics Press, Washington, D.C., 1986. Contains dielectric properties data, especially relating to spacecraft charging. Worth obtaining and reading.

APPENDIX A
Nomenclature

A.1 CONSTANTS AND MEASUREMENT UNITS

A	ampere (unit of current)
AU	astronomical unit (Earth-to-Sun distance, ~150,000,000 km)
C	coulomb
cm	centimeter
dB	decibel
deg	degree
e/cm^2	electrons per square centimeter
eV	electron volt
F	farad (measure of electrical capacitance)
ft	foot
GHz	gigahertz
h	hour
Hz	hertz, unit of frequency (1 cycle per second)
in.	inch
J	joule (unit of energy)
K	kelvin
kÅ	kiloangstrom
keV	kiloelectron volt (10^3 eV)
kg	kilogram
km	kilometer
kΩ	kilohm (10^3 Ω)
kV	kilovolt (10^3 V)
m	meter, mass, milli
M	mega, million
mA	milliampere (10^{-3} A)
MeV	million electron volts
mho	unit of electrical conductance, the reciprocal of resistance in Ω, also known as the siemens (S)
MHz	megahertz (frequency, 10^6 Hz)

Guide to Mitigating Spacecraft Charging Effects, First Edition. Henry B. Garrett, Albert C. Whittlesey.

mil	one-thousandth of an inch = 0.001in. = 0.0254 mm [*Note* Although the mil is not an SI unit of measure, it is a standard unit of measure (unconverted to SI units) often used in this discipline to describe material thickness].
min	minute
mJ	millijoulc (10^{-3} J)
mm	millimeter
mΩ	milliohm (10^{-3} Ω)
MΩ	megohm (10^{6} Ω)
mV	millivolt
nA	nanoampere (10^{-9} A)
nC	nanocoulomb (10^{-9} C)
nF	nanofarad (10^{-9} F)
nm	nanometer (10^{-9} m)
ns	nanosecond (10^{-9} s)
nT	nanotesla (10^{-9} tesla), magnetic field unit
ohm (Ω)	unit of electrical resistance
pA	picoampere (10^{-12} A)
pF	picofarad (10^{-12} F)
q, Q	charge (coulombs)
R	resistance (ohms, Ω)
R_e	radius compared to Earth (1 R_e ~6378.136 km)
R_j	radius compared to Jupiter (1 R_j ~7.1492 × 10^4 km)
R_s	radius compared to Saturn (1 R_s ~6.0268 × 10^4 km)
s/d	seconds per day (86,400)
s/h	seconds per hour (3600)
Sr	steradian
μC	microcoulomb (10^{-6} C)
μF	microfarad (10^{-6} F)
μJ	microjoule (10^{-6} J)
μm	micrometer (10^{-6} m)
μs	microsecond (10^{-6} s)
μW	microwatt (10^{-6} W)
V	volt
V	voltage
W	watt

A.2 ACRONYMS AND ABBREVIATIONS

1-D	one-dimensional
2-D	two-dimensional
3-D	three-dimensional
ACE	Advanced Composition Explorer

ac, AC	alternating current
ACR	anomalous cosmic ray
ADEOS-II	Advanced Earth Observing Satellite-II; Japanese satellite (802.92 km, 98.62°, 101 min), Dec. 2002–Oct. 2003
AE8	NASA Space Radiation Model for Trapped Electrons
AF	Air Force
AFB	Air Force Base
AFGL	Air Force Geophysics Laboratory
AFRL	Air Force Research Laboratory
AIAA	American Institute of Aeronautics and Astronautics
ALT	altitude
AP8	NASA Space Radiation Model for Trapped Protons
ASTAR	Web databases used to calculate stopping powers, ranges, and related quantities, for electrons, protons, and helium ions
ASTM	American Society for Testing and Materials
ATS	Applications Technology Satellite (ATS-5 and ATS-6) geostationary satellites
CEASE	Compact Environmental Anomaly Sensor
CME	coronal mass ejection
CMOS	complementary metal-oxide semiconductor
CNES	Centre National d'Etudes Spatiales, the French space agency
CPA	Charged Particle Analyzer
CPE	charged particle environment
CREME96	Cosmic Ray Effects on MicroElectronics 1996 (environmental code)
CRRES	Combined Release and Radiation Effects Satellite
CRRESELE	CRRES electron flux energy spectrum environmental code
CRRESPRO	CRRES proton flux energy spectrum environmental code
CRRESRAD	CRRES dose versus depth environmental code
CTS	Communications Technology Satellite
DC, dc	direct current (zero frequency)
DDD	displacement damage dose
DERA	Defense Evaluation and Research Agency
DESP	Space Environment Department (France)
DICTAT	DERA Internal Charging Threat Analysis Tool
div	division
DMSP	Defense Meteorological Satellite Program (800 km; 99°, 110 min)
DoD	Department of Defense
DS-1	*Deep Space 1* (spacecraft)
DynaPAC	Dynamic Plasma Analysis Code
ECSS	European Cooperation on Space Standardization
e.g.	for example
EGS4	Monte Carlo transport code
EMC	electromagnetic compatibility

EMI	electromagnetic interference
EMP	electromagnetic pulse
ESA	European Space Agency
ESD	electrostatic discharge
ESTAR	Web databases used to calculate stopping powers, ranges, and related quantities, for electrons, protons, and helium ions
etc.	and so forth
EURECA	European Retrievable Carrier
EVA	extravehicular activity
EUV	extreme ultraviolet
EWB	Environmental Work Bench
FEP	fluorinated ethylene propylene (FEP; Teflon®)
FLUMIC	Flux Model for Internal Charging
FR4	flame retardant 4, common printed circuit board material
GaAs	gallium arsenide
Galileo	European plan for a GPS-like system of satellites [23,222 km (14,429 mi) altitude, 56° inclination]
Galileo	NASA spacecraft sent to Jupiter, launched October 18, 1989, ended deliberately September 21, 2003
GCR	galactic cosmic ray
Geant4	particle transport code, the European counterpart to MCNPX
GEO	geosynchronous Earth orbit [about 35,786 km (22,236 mi) altitude, 24-h period]
GEOSTA	geostationary
GHz	gigahertz (10^9 Hz)
Giove A, B	MEO precursor to European Galileo GPS satellite constellation; launched Dec. 28, 2005, and April 27, 2008, respectively
GIRE	Galileo Interim Radiation Electron Model
GOES	Geosynchronous Operational Environmental Satellite
GPS	Global Positioning Satellite [constellation, 20,100 km (12,490 mi), 55°, 718 min]
GRC	Glenn Research Center [formerly Lewis Research Center LeRC]
GSE	Ground Support Equipment (or SE)
GSFC	Goddard Space Flight Center
GUI	graphical user interface
H^+	hydrogen ion
HBM	human body model
HDBK	handbook
HEO	highly elliptical orbit, used synonymously for Molniya
IC	integrated circuit
IDM	Internal Discharge Monitor (flown on CRRES)
i.e.	that is
IEEE	Institute of Electrical and Electronics Engineers

IESD	internal electrostatic discharge
INTELSAT	International Telecommunications Satellite Organization
IR	infrared
ISO	International Standards Organization
ISPICE	computer transient circuit analysis program; first commercial version of SPICE
ISS	International Space Station [~390 km (~240 mi) altitude (varies), 51.6°, 92 min]
ISTP	International Solar-Terrestrial Physics
ITAR	international traffic-in-arms regulations (restricted access to some information)
ITO	indium tin oxide
ITS	Integrated TIGER Series
I/V	current versus voltage
JAXA	Japanese Space Agency
JPL	Jet Propulsion Laboratory
LANL	Los Alamos National Laboratory
LAT	latitude
LED	light-emitting diode
LEM	lumped-element model
LEO	low-Earth orbit [about 200 to 2000 km (124 to 1240 mi) altitude; e.g., 657 km, 1.5-h period]
LET	linear energy transfer
LeRC	Lewis Research Center [now Glenn Research Center (GRC)]
LNA	low-noise amplifier
MCNP	Monte Carlo N-Particle Transport Code
MCNPE	version of MCNP modified to include transport of electrons
MCNPX	Monte Carlo N-Particle eXtended Transport Code
MEO	medium Earth orbit [about 2000 to 25,000 km (1240 to 15,500 mi) altitude ~6 h period]
MIL	military
MLI	multilayer insulation (thermal blanket)
Molniya	elliptical orbit (apogee ~39,300 km, perigee 538 km, 11.8-h period, ~63.2° inclination)
MPA	Magnetospheric Plasma Analyzer
MSFC	Marshall Space Flight Center
MUSCAT	Multi Utility Spacecraft Charging Analysis Tool
NASA	National Aeronautics and Space Administration
NASCAP	NASA Charging Analyzer Program, historic and generic
NASCAP/GEO	NASA Spacecraft Charging Analyzer Program for Geosynchronous Orbit (replaced by Nascap-2k)
NASCAP/LEO	NASA Spacecraft Charging Analyzer Program for Low Earth Orbit
Nascap-2k	latest version of NASCAP (as of 2011)
NGST	Northrop-Grumman Space Technology

NOAA	National Oceanic and Atmospheric Administration
NOVICE	charged-particle radiation transport code
NSSDC	National Space Science Data Center
NUMIT	numerical model for estimating charging in dielectrics
ONERA	Office National d'Etudes et Recherches Aérospatiales, the French national aerospace research center
OSR	optical solar reflector
p	proton
PEO	polar Earth orbit [$\sim 80^\circ$ or higher inclination, 700 to 1000 km (435 to 620 mi) altitude $\sim$100-min period]
PET	Proton/Electron Telescope
photo	photon–emitted particles (e.g., photoelectrons)
PIC	particle in cell
PIX, PIX-II	Plasma Interactions Experiment
POES	Polar Operational Environmental Satellite [$\sim 80^\circ$ or higher inclination 700 to 1000 km (435 to 620 mi) altitude]
POLAR	Potential of Large Objects in the Auroral Region (NASCAP model)
PSTAR	Web databases used to calculate stopping powers, ranges, and related quantities, for electrons, protons, and helium ions
PTFE	polytetrafluoroethylene (Teflon®)
QA	quality assurance
RC	resistor–capacitor
RF	radio frequency
RIC	radiation-induced conductivity
RP	reference publication
RSICC	Radiation Shielding Information Computational Center
RTV	room-temperature vulcanized (adhesive)
SAIC	Science Applications International Corporation
SAMPEX	Solar, Anomalous, and Magnetospheric Particle Explorer
SAMPIE	Solar Array Module Plasma Interactions Experiment
SATRAD	Saturn Radiation model
SCATHA	Spacecraft Charging at High Altitudes Satellite [1979–1986, 28,000 × 42,000 km (17,400 to 26,100 mi) altitude, 8.3° inclination]
SCR	silicon-controlled rectifier
SCTC	Space Communications Technology Center
SE	support equipment (nonflight hardware) (or GSE)
Sec	secondary emission
SEE	single-event effect
SEE	(NASA) Space Environment Effects Program
SEMCAP	Specification and Electromagnetic Compatibility Program
SEP	solar energetic particle (or proton) event
SEU	single-event upset
SHIELDOSE	charged-particle radiation transport code

Si	silicon
SI	Système Internationale (based on the older metric system of measurement)
SOHO	Solar and Heliospheric Observatory
SOPA	Synchronous Orbit Particle Analyzer (Los Alamos)
SPE	solar proton event
SPENVIS	Space Environment Information System
SPICE	Simulation Program for Integrated Circuit Emphasis, a computer transient circuit analysis program
SPINE	Spacecraft Plasma Interactive Network
SPIS	Spacecraft Plasma Interactive System
sqrt	square root
SSO	semisynchronous orbit, ~20,000 km (12,400 mi), 12-h period
STD	standard
TID	total ionizing dose
TP	technical publication
TRACE	Transition Region and Coronal Explorer
TRIM	radiation transport code
TRW	TRW Incorporated (now Northrop-Grumman)
TSS-1R	Tethered Satellite System—First Re-Flight
UCSD	University of California at San Diego
USA	United States of America
USAF	U.S. Air Force
UV	ultraviolet
VDA	vacuum-deposited aluminum
WDC	World Data Center (NOAA)
ZOT	zinc orthotitanate paint

A.3 DEFINED TERMS

Ap index	Daily worldwide geomagnetic activity index.
Auroral Zone	Geomagnetic latitudes between ~60 and 70° north/south, where auroras are present.
Blow-off	The effect of an ESD when material and charge are expelled by an ESD.
Bonding	As used for ESD and electromagnetic compatibility (EMC), attaching something electrically to spacecraft chassis or electrically attaching conductive chassis parts to each other, as distinct from a deliberate current-carrying path; sometimes also called grounding, but generally there is not a problem in context.
Buried charging	Refers to charging internal to a spacecraft, often meaning within dielectrics, but possibly in

	ungrounded (floating) metals. The authors prefer the term *internal charging*.
Conductor	For the purpose of this spacecraft charging document, a conductor is a material that is used for carrying current or is similarly conductive and acting as part of a shield or ground plane structure. Copper and aluminum are typical conductors. (*See also* Insulator; Dielectric; ESD/static-conductive).
Debye length	Characteristic distance (λ_D) in a plasma over which a plasma screens out (e.g., reduces) the electric field by $1/e$.
Deep dielectric charging	Charging internal to dielectrics caused by energetic electrons. (*See also* Buried charging.) The authors prefer the term *internal charging* unless referring specifically to dielectrics.
Dielectric	For the purpose of spacecraft charging, a dielectric is a resistive material that may be synonymous with *insulator*. This document suggests that dielectrics have a bulk resistivity of $>10^{10}\ \Omega \cdot \text{cm}$ or a surface resistivity of $>10^9\ \Omega/\text{square}$. (*See also* Insulator; Conductor; ESD/static-conductive.)
ESD/static-conductive	For the purpose of spacecraft charging, an ESD/static-conductive material is one that is adequately conductive to conduct any space plasma charges to ground so that the charging effects have minimal or no impact on spacecraft operations. These are partially resistive materials that are neither conductors nor insulators. There is not an official definition in this document, but an approximate range of resistivity for ESD/static conductive materials is less than $10^8\ \Omega/\text{square}$ for thin materials and $10^7\ \Omega \cdot \text{cm}$ for bulk materials. And it must be grounded properly to be useful for mitigation of spacecraft charging. (*See also* Dielectric; Conductor; Isolation.)
Faraday cage	A completely enclosed metallic container; an electromagnetically shielded enclosure.
Floating	A conductor is floating if it is ungrounded or has no defined reference to chassis. (*See also* Referenced.)
Geostationary	A geosynchronous orbit directly above the Earth's equator (0° latitude), with a period equal to the

	Earth's rotational period and an orbital eccentricity of approximately zero. At a fixed position in the sky, an object in a geostationary orbit appears motionless to ground observers.
Geosynchronous	A circular orbit in the equatorial plane of Earth at stationary orbit, ~35,768 km (22,225 mi) altitude that matches the Earth's sidereal rotation period. The synchronization of rotation and orbital period means that for an observer on the surface of the Earth, the satellite appears to hover constantly over the same meridian (north–south line) on the surface, moving in a slow oscillation alternately north and south with a period of one day, so it returns to exactly the same place in the sky at exactly the same time each day.
Ground	A connection to a zero-volt reference point (ground), often the chassis. *Note:* Bonding is used almost exclusively as a connection to chassis for other purposes, such as space charge bleed-off, shield terminations, or fault current paths. Structure is used as ground for both bonding and circuit zero volt referencing, so the term *structure ground* is often used interchangeably with the term *signal ground*. For this reason, be careful when using the word *ground*.
Insulator	For the purpose of spacecraft charging, an insulator is a highly resistive material that does not have adequate conductivity to discharge charge accumulation coming from the environment. There is not an official definition in this document, but an approximate range of resistivity for insulators is greater than 10^9 Ω/square for thin materials and 10^8 Ω · cm for bulk materials. (*See also* Dielectric; Conductor; ESD/static-conductive.)
Internal charging	The buildup of charge on the interior parts of a spacecraft from higher-energy particles.
L1–L5	Lagrange/libration points (astronomical). For a third body, locations of orbital positions requiring minimum energy maintenance with respect to two other (larger) bodies.
Molniya	An elliptical orbit with an apogee of ~39,300 km (~24,400 mi), a perigee of 538 km (334 mi),

	an 11.8-h period, and a ~63.2° inclination. Molniya orbits are named after a series of Soviet/Russian *Molniya* (Russian for "lightning") communications satellites which have used this type of orbit since the mid-1960s.
ohm per square	A measure of surface resistivity. The resistance of a flat relatively thin sheet of the material, measured from one edge of a square section to the opposite edge. Properly, units are ohms.
Referenced	Not ungrounded, meaning that there is a defined path to ground even if the referenced item is not at ground. For example, the +28-V power line is not grounded, but it is referenced to ground, and thus it is not floating. It cannot accumulate stray charges.
Spacecraft charging	The buildup of charge in and on spacecraft materials; a significant phenomenon for spacecraft in certain Earth and other planetary environments.
Triple junction point	In this document, refers to a place in solar arrays where a dielectric, a conductor, and space all meet at one point. Intense electric fields may exist and cause ESDs at solar array triple junction points.
Victim	Any part, component, subsystem, or element of a spacecraft that can be affected adversely by an arc discharge (or field effects in the case of some science instruments).

A.4 VARIABLES

(typical units unless specified otherwise)

BS	backscattered
C	capacitance
CPH	photoelectron current
dE/E	energy channel width (dE) expressed as fraction of nominal median energy (E) for channel
E	electric (fields), energy, or East
H-field	magnetic field (common usage)
i	differential angular intensity (or flux) (example: ions/cm$^2 \cdot$s$\cdot$Sr$\cdot$keV)
I	integral angular intensity (or flux) (example: electrons/cm$^2 \cdot$ s $\cdot$ Sr)
I	current (A)

I_{pk}	peak current (A)
j	omnidirectional differential flux (example: electrons/ $cm^2 \cdot s \cdot MeV$)
J	omnidirectional integral flux (example: electrons/$cm^2 \cdot s$)
J	current per unit area (A/cm^2)
N_E	electron number density
N_I	ion number density
R	resistance (Ω)
R_E	density of electron plasma environment
R_I	density of ion plasma environment
T	temperature
T_E	temperature for electron plasma environment
T_I	temperature for ion plasma environment
TP	technical publication
T_R	discharge current risetime
v	velocity
V_b	voltage breakdown; breakdown voltage
V_c	corotation velocity of specified region
V_{sw}	solar wind bulk velocity
V_{th}	solar wind thermal velocity
yr	year

A.5 SYMBOLS

$\sim$	approximately
$\circ$	degree
$>$	greater than
$<$	less than
$\pm$	plus or minus
d	day
e	electron (charge = 1.6022×10^{-19} coulomb)
ε	total permittivity $\varepsilon = \varepsilon_0 \times \varepsilon_r$, dielectric constant
ε_0	free-space permittivity (= 8.85×10^{-12} F/m)
ε_r	relative permittivity
g	gram
H	magnetic field (or B in free space)
l	length
μ	mu, or micro, representing a factor of 10^{-6}
ρ	rho (volume resistivity) (ohm · m or ohm · cm, Ω · m, or Ω · cm)
ρ_s	rho–sub s (surface resistivity) (SI unit: ohm; more commonly, "ohms per square")
s	second

S	siemens (reciprocal of resistance, also mho) ($1/R$)
σ	sigma [conductivity; units: $(\Omega \cdot \text{cm})^{-1}$)
t	time, thickness
Ω	resistance (in ohms)

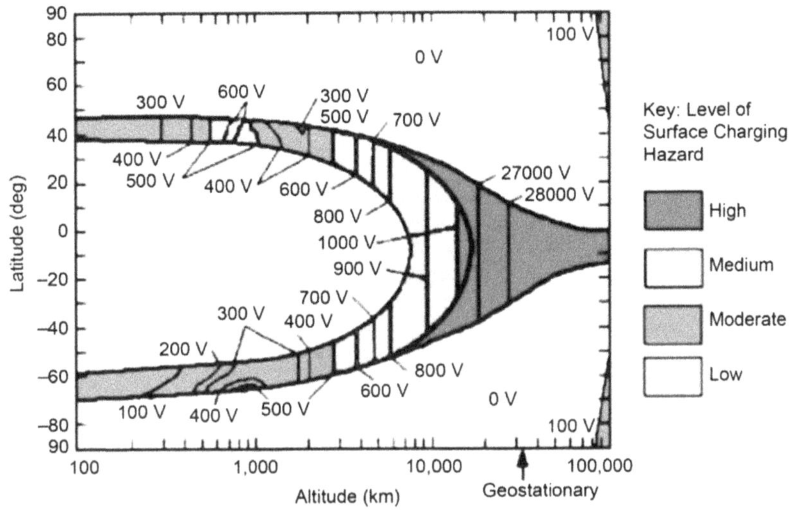

Fig. 1-1. Earth regimes of concern for on-orbit surface charging hazards for spacecraft passing through the latitude and altitude indicated. See Whittlesey et al. [9] for an alternative reference with the "Wishbone" chart. (From [8].)

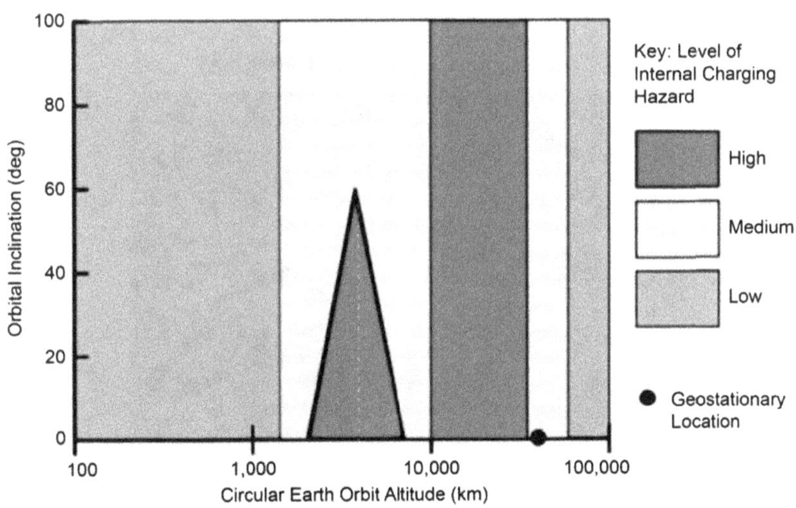

Fig. 1-2. Earth regimes of concern for on-orbit internal charging hazards for spacecraft with circular orbits.

Guide to Mitigating Spacecraft Charging Effects, First Edition. Henry B. Garrett, Albert C. Whittlesey.

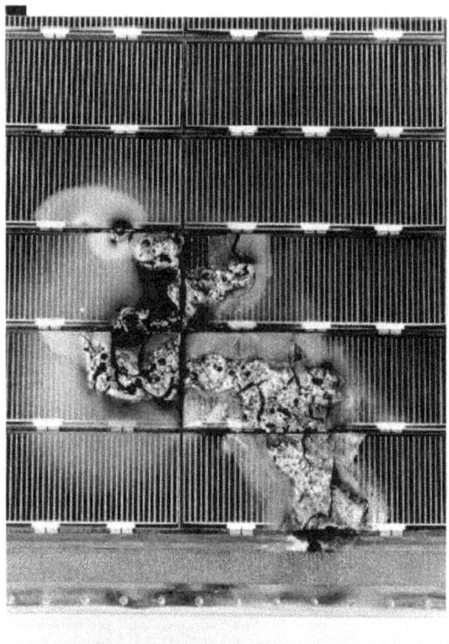

(a) Failure caused by in-flight arcing.

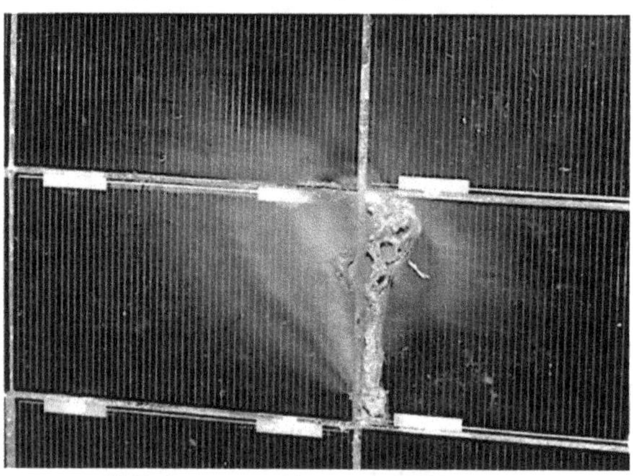

(b) Failure caused by ground ESD arcing.

Fig. 3-2. Examples of solar array failures caused by (a) in-flight arcing (from [15]) and (b) ground ESD arcing.

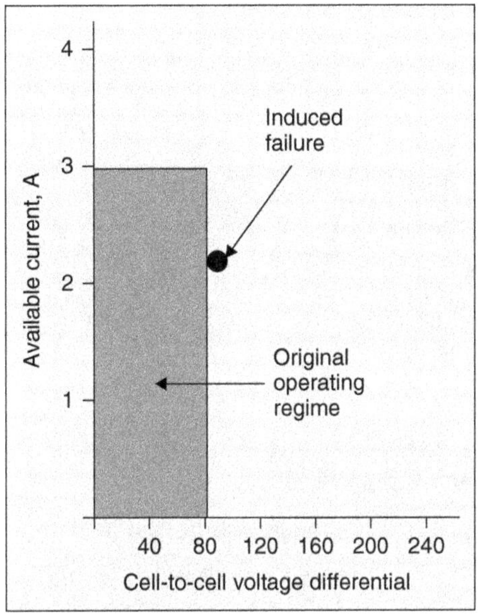

Fig. 3-3. Measured gallium arsenide (GaAs) coupon I/V failure threshold (From [3].)

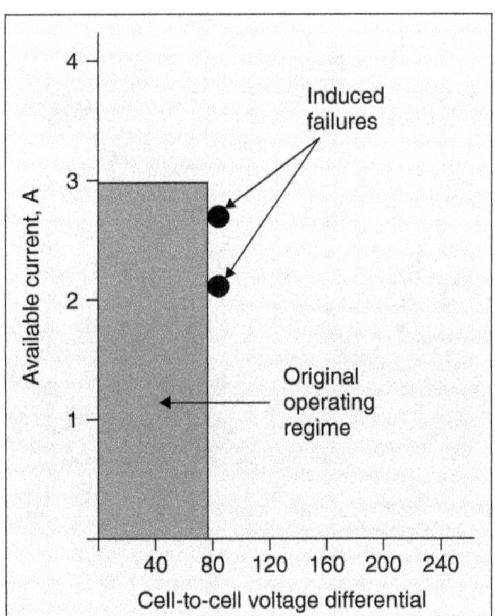

Fig. 3-4. Measured silicon (Si) coupon I/V failure threshold (From [3].)

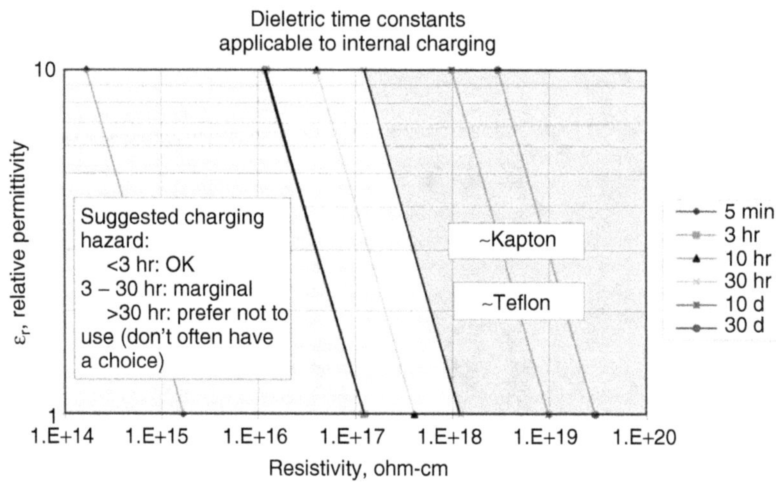

Fig. 6-1. Safe, intermediate, and possibly hazardous dielectric materials based on resistivity and dielectric constant and resulting time constant. (The Kapton® and Teflon® boxes illustrate the uncertainty range for space applications; see the text.)

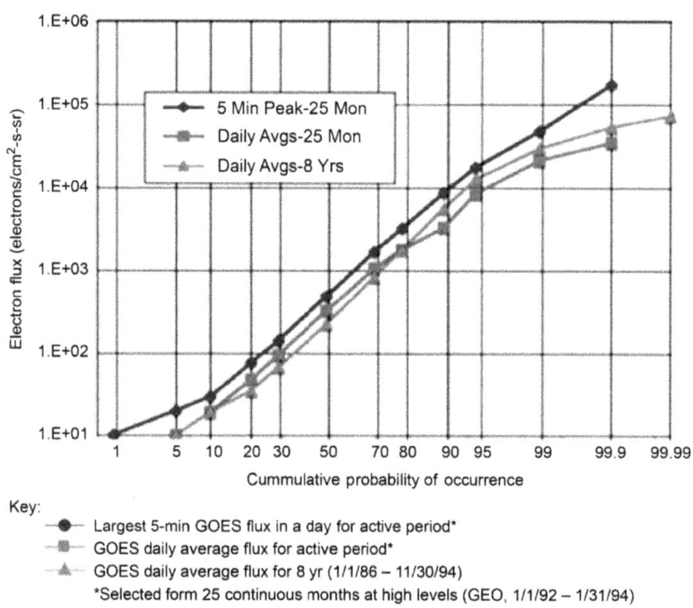

Fig. B-7. Cumulative probability of occurrence of GOES-7 $E > 2$ MeV electron fluxes for several different assumptions.

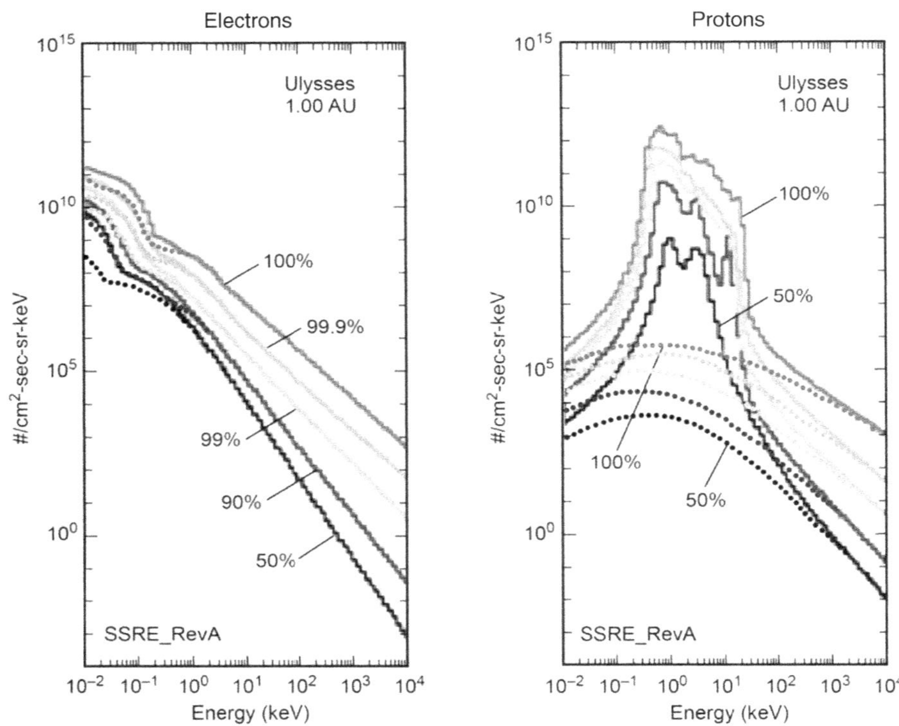

Fig. B-10. Solar wind parameters for an interplanetary shock and high- and low-speed stream versus time as measured by the SOHO spacecraft. Based on the CELIAS/MTOF experiment on the Solar Heliospheric Observatory (SOHO) spacecraft. SOHO is a joint mission of the European Space Agency and the National Aeronautics and Space Administration. (From [22]; SOHO CELIAS/Proton Monitor data, courtesy of University of Maryland).

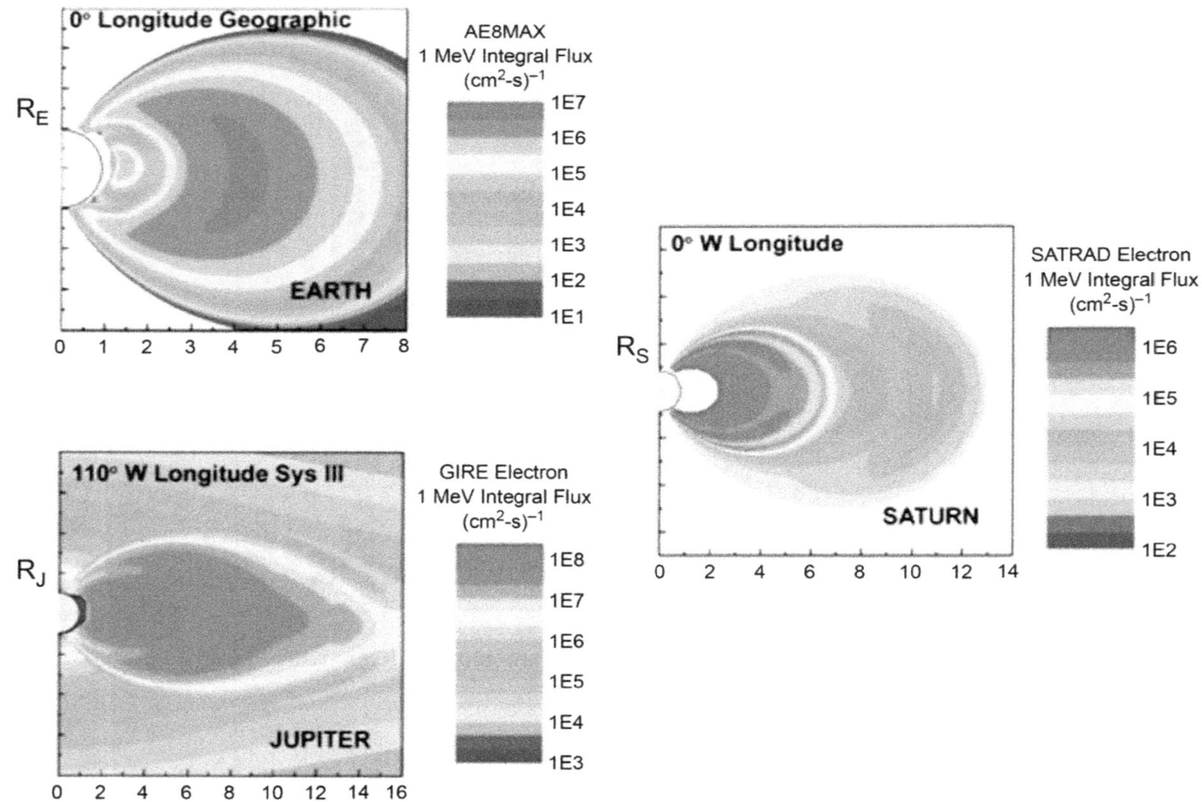

Fig. B-11. 1-MeV electron omnidirectional flux contours for Earth, Jupiter, and Saturn (models used: AE8Max, GIRE, and SATRAD). (Courtesy of Insoo Jun, personal communication.)

APPENDIX B
The Space Environment

B.1 INTRODUCTION TO SPACE ENVIRONMENTS

This appendix is intended to supplement the material presented in Chapter 2. It presents in more detail, for the interested reader, many of the concepts introduced in Chapter 2.

B.1.1 Quantitative Representations of the Space Environment

Earth's plasma is described properly in terms of phase space density or distribution function. Space plasmas can be described most simply in terms of the Maxwell–Boltzmann distribution. As this representation lends itself to efficient manipulation when carrying out charging calculations, it is often the preferred way of describing plasmas. The Maxwell–Boltzmann distribution F_i is given by

$$F_i(v) = \left[n_i \left(\frac{m_i}{2\pi k T_i} \right)^{3/2} \right] \exp\left(\frac{-m_i v^2}{2kT_i} \right) \tag{B-1}$$

where

n_i = number density of species i
m_i = mass of species i
k = Boltzmann constant
T_i = characteristic temperature of species i
v = velocity
F_i = distribution function of species i

Unfortunately, the space plasma environment is seldom a Maxwell–Boltzmann distribution. However, given the actual plasma distribution function, it is possible to define (irrespective of whether the plasma is Maxwell–Boltzmann or not) moments of the particle distribution that reveal characteristics of its shape.

Guide to Mitigating Spacecraft Charging Effects, First Edition. Henry B. Garrett, Albert C. Whittlesey.

In most cases, these moments can then be used to determine an approximate Maxwell–Boltzmann distribution. The first four of these characteristic moments are

$$<ND_i> = 4\pi \int_0^\infty (v^0) F_i v^2 \, dv \tag{B-2}$$

$$<NF_i> = \int_0^\infty (v^1) F_i v^2 \, dv \tag{B-3}$$

$$<ED_i> = \frac{4\pi m_i}{2} \int_0^\infty (v^2) F_i v^2 \, dv \tag{B-4}$$

$$<EF_i> = \frac{m_i}{2} \int_0^\infty (v^3) F_i v^2 \, dv \tag{B-5}$$

where

$<ND_i>$ = number density of species i
$<NF_i>$ = number flux of species i
$<ED_i>$ = energy density of species i
$<EF_i>$ = energy flux of species i

For the Maxwell–Boltzmann distribution of Eq. (B.1-1), these assume the following values:

$$<ND_i> = n_i \tag{B-6}$$

$$<NF_i> = \frac{n_i}{2\pi}\left(\frac{2kT_i}{\pi m_i}\right)^{1/2} \tag{B-7}$$

$$<ED_i> = \frac{3}{2} n_i k T_i \tag{B-8}$$

$$<EF_i> = \frac{m_i n_i}{2}\left(\frac{2kT_i}{\pi m_i}\right)^{3/2} \tag{B-9}$$

It is often easier to measure the moments (e.g., number flux, of the plasma distribution function) than the actual distribution function in terms of energy or the temperature. This is particularly true for space plasmas where the concept of temperature is not well defined. As an illustration, from the first four moments, two definitions of the plasma temperature consistent with a Maxwell–Boltzmann distribution are possible:

$$T_{av} = \frac{2<ED>}{3<ND>} \tag{B-10}$$

$$T_{rms} = \frac{<EF>}{2<NF>} \tag{B-11}$$

For a true Maxwell–Boltzmann plasma, these quantities would be equal; for actual plasmas, T_{rms} is usually greater than T_{av}. Even so, experience has shown that a representation in terms of two Maxwell–Boltzmann distributions is, in fact, a better mathematical representation of the space plasma than a single Maxwellian. That is, the plasma distribution for a single species can be represented by

$$F_2(v) = \left(\frac{m}{2\pi k}\right)^{3/2} \left[\frac{N_1}{(T_1)^{3/2}} \times \exp\left(\frac{-mv^2}{2kT_1}\right) + \frac{N_2}{(T_2)^{3/2}} \exp\left(\frac{-mv^2}{2kT_2}\right)\right] \quad \text{(B-12)}$$

where

N_1 = number density for population 1
T_1 = temperature for population 1
N_2 = number density for population 2
T_2 = temperature for population 2

In most cases, this representation fits the data quite adequately over the energy range of importance to spacecraft surface charging: ~1 eV to 100 keV. Further, it is very simple to derive N_1, T_1, N_2, and T_2 directly from the four moments so that a consistent mathematical representation of the plasma can be established that incorporates the simplicity of the Maxwell–Boltzmann representation while maintaining a physically reasonable picture of the plasma. The distinction between T_{av}, T_{rms}, T_1, and T_2 must be kept in mind, however, whenever reference is made to a Maxwell–Boltzmann distribution, as this is, at best, only an approximation to the actual plasma environment.

Although the Maxwell–Boltzmann distribution can be used to represent the high-energy electron environment for internal charging, it is typically not as useful as it is for surface charging calculations. More typically, the electron environment above about 100 keV approaches a functional form represented by a power law or the more complex kappa distribution, which better represents the nonthermal tail in the electron distribution at higher energies. For example, if a power law distribution A_0E^{-x} is assumed for $i(E)$, the differential intensity (also often called *flux*), the integral intensity (E) would give

$$I(E) = -\int_E^{\infty} i(E)\,dE = -\frac{(A_0E^{1-X})}{1-X} \quad \text{(B-13)}$$

where

$i(E)$ = - $dI(E)/dE$ = differential angular intensity (or flux) = particles per unit area per unit energy per unit of solid angle at energy E (example: $n/\text{cm}^2 \cdot s \cdot Sr \cdot \text{keV}$)
$I(E)$ = integral (over energy) angular intensity (or flux) = particles per unit area per unit of solid angle from energy E to infinity (example: $n/\text{cm}^2 \cdot s \cdot Sr$)
E = energy of particle
A_0, X = constants

The omnidirectional fluxes are then given by

$$j(E) = \int_0^{\pi} d\alpha \int_0^{2\pi} i(E) \sin\alpha \, d\phi \tag{B-14}$$

$$J(E) = \int_0^{\pi} d\alpha \int_0^{2\pi} I(E) \sin\alpha \, d\phi \tag{B-15}$$

where

$j(E)$ = omnidirectional differential flux = particles per unit area per unit energy integrated over 4π steradians at energy E (example: $n/\text{cm}^2 \cdot s \cdot \text{MeV}$)

$J(E)$ = omnidirectional integral flux = particles per unit area over 4π steradians from energy E to infinity (example: $n/\text{cm}^2 \cdot s$)

α = particle pitch angle (radians) for particles in a magnetic field or, in the absence of a magnetic field, the angle relative to the normal to a surface

Some publications, including NASA's AE8/AP8 family of radiation models, use the term *omnidirectional integral flux* as defined above, which implies an isotropic (uniform in all directions) particle flux. This is our J or the omnidirectional integral flux. Other publications report intensity (flux) per steradian (or our I with units of $n/\text{cm}^2 \cdot \text{s} \cdot \text{Sr}$). Assuming an isotropic plasma (a common simplifying assumption), the two are related by

$$J = 4\pi I \tag{B-16}$$

Similarly, after multiplying by charge, q, and converting from charge/s to amperes, the net current per unit area, J, to a flat surface for an isotropic flux, when integrated over an angle [Eq. (B.1-15)] can be shown to be

$$\text{J} = \pi q I \qquad \text{A/cm}^2 \tag{B-17}$$

The reduction of $\frac{1}{4}$ is due to two factors. The first $\frac{1}{2}$ is because the current to a surface comes from only one side of the surface. The second $\frac{1}{2}$ is the average value of current due to the integral over the angle for nonnormal incidence. If the flux is not isotropic, these simple calculations must be redone for the actual angular distribution.

[*Note:* To avoid confusion, in the rest of the book the current to a spacecraft is defined as I, where I = J × (collection area).]

The preceding is true for the fluxes and currents impacting the surface. For penetration calculations, the geometry of the shielding must be considered carefully in estimating the fluxes in a material or inside the shielding. For example, the nonnormally incident electrons cannot penetrate as deep as normally incident electrons because of the longer path length through the shielding to a given point. The difference depends on the depth and on the spectrum of the electrons; accurate calculations require specialized codes, discussed in Appendix C.

B.1.2 Data Sources

In the following subsections we list briefly the satellites and sources from which environmental data can be obtained. Note that there are problems in attempting to obtain calibrated particle data from space. Energetic electron detector data are, as an example, sometimes affected by the presence of energetic protons that generate secondary electrons during their passage through the detector. Detectors may degrade and become less efficient over time or may not even be initially calibrated over all energy ranges. View factors and orientation relative to the magnetic field also contribute to uncertainties in the count rate-to-flux conversion. Despite these concerns, the errors are usually small enough to permit the data to be used in estimating charging, at least for engineering purposes.

B.1.2.1 ATS-5, ATS-6. A major source of data on the geosynchronous plasma environment has been the University of California at San Diego (UCSD) low-energy plasma detectors on the NASA geosynchronous satellites ATS-5 and ATS-6. In particular, data were taken for electrons and ions (assumed to be protons) in 62 energy channels. For ATS-5, at a longitude of ~225°E, spectra were taken every 20 s in 112% (dE/E) energy intervals from 51 eV to 51 keV. For ATS-6, at a longitude of ~266°E, spectra were taken every 15 s in 113% dE/E intervals from 1 eV to 81 keV. The data are available from the National Space Science Data Center (NSSDC) in 10-min average bins for 50 days between 1969 and 1970 for ATS-5 and 10-min bins for 45 days between 1974 and 1976 for ATS-6. The data are in the form of observation time, spacecraft coordinates, and the four moments of the electron and ion distribution functions. These data were analyzed extensively in papers by Garrett and colleagues [1–3]. They, along with data from SCATHA, represented the primary source of statistical data on the geosynchronous orbit until recent studies of the Los Alamos National Laboratory (LANL) instruments (Section B.1.2.4). An additional 10 days of data from ATS-6 are also available for a unique period (September 14–25, 1976), during which the ATS-6 spacecraft passed by the LANL Charged Particle Analyzer (CPA) instrument on another geosynchronous spacecraft, allowing careful cross-calibration of the particle instruments. Some descriptions of these data appear in reference [4]. Reference [5] provides an excellent summary of Earth's space plasma environments that sets the context for these observations.

B.1.2.2 SCATHA. Launched in 1979, the SCATHA satellite is another major source of spacecraft charging data. In addition to numerous experiments for measuring and controlling spacecraft charging, SCATHA measured the space environment between 5.5 and 7.7 R_e for a number of years. Of particular interest to environmental studies are the Air Force Geophysics Laboratory (AFGL) SC5 Rapid Scan Particle Detector which measured the electron and ion environments at 1-s intervals over the range 50 eV to 0.5 MeV, and the UCSD SC9 Low-Energy Plasma Detector which measured the electron and ion plasma every 0.25 s at energies of 1 eV to 81 keV, the instrument being a near-duplicate of the

ATS-5 and ATS-6 instruments. As in the case of these two spacecraft, the data were analyzed extensively by Mullen and colleagues to return similar statistical results that can be compared to the ATS-5 and ATS-6 findings [6–9]. The data are available in the referenced documents and some through the NSSDC.

B.1.2.3 GOES. The most readily available data on the high-energy particle environments are those from the National Oceanic and Atmospheric Administration (NOAA) Geosynchronous Operational Environmental Satellite (GOES) series of spacecraft at geosynchronous orbit. The data of interest here consist primarily of $E > 2$ MeV electron fluxes expressed in $e/cm^2 \cdot s \cdot Sr$. Starting with GOES 8, data are also available for the $E > 600$ keV electron environment. Data from at least early 1986 to the present are readily available. GOES satellites are generally positioned over the U.S. east and west coasts, but their exact positions have varied over the years. Contact Dan Wilkinson, phone 303-497-6137. Data are available in near real time over the web at http://ngdc.noaa.gov/; click on "Space Weather & Solar Events," then click on "Satellite Data Services: GOES SEM" and select from various options. Alternatively, at the home page, look at various selection options. Go to URL http://www.swpc.noaa.gov/today.html for the last 3 days of GOES space weather data.

B.1.2.4 Los Alamos Detectors. Detectors on board various Department of Defense geosynchronous spacecraft provided by the LANL have been in service since the 1970s. Higher-energy channels are referred to as CPA or, currently, the SOPA experiments. The data cover a wide energy range (e.g., from $E > 30$ eV to $E > 5$ MeV for electrons) and are available from 1976 through 2005. The data are well calibrated and provide a more detailed snapshot of the environment than the GOES data but have not been as readily available. Recent papers presenting the Los Alamos data are references [10] and [11]. Contact Michelle Thomsen, phone 506-667-1210, or Geoff Reeves, phone 505-665-3877. The LANL data Web site can be accessed at http://leadbelly.lanl.gov/. Historical to current energetic particle data can be obtained at that site.

In addition to SOPA, since 1989, LANL has been accumulating high-quality measurements of electron and proton energy flux spectra from 1 eV to 40 keV from Magnetospheric Plasma Analyzer (MPA) instruments aboard a series of geosynchronous spacecraft. These data not only characterize the plasma but can also be used to infer the potential (relative to plasma) of the instrument ground and the presence of differential charging. From the raw data, spin-angle-averaged flux spectra, spacecraft potential, and various moments are computed. The density and temperature moments should be used cautiously with a full understanding of how they are computed (see reference [12] for details of the data analysis). Reference [13] provides statistics on the electrons and ions over a full solar cycle along with detailed spectra. Spectrograms and moments can be obtained from Michelle Thomsen at mthomsen@lanl.gov, as can further information and specific data.

B.1.2.5 CRRES. Launched in 1990, the Combined Release and Radiation Effects Satellite (CRRES) spacecraft provided the most accurate and detailed measurements of Earth's radiation belts in many decades. A landmark in internal charging (it carried the first experiment specifically designed to study internal charging), it provided extensive data on the location and occurrence of IESDs throughout the magnetosphere. CRRES was launched into an eccentric, 18° inclination orbit that took it from below the Van Allen belts out to geosynchronous orbit. It had an orbital period of 10 h and measured electrons from a few eV to 10 MeV. The primary data are from July 25, 1990 to October 1991, and include extensive measurements of internal arcing rates in addition to the radiation data. These data and related software codes may be obtained via a Google search of AF-GEOSPACE; use link fact sheets: AF-GEOSPACE; a software request form is provided.

B.1.2.6 Solar, Anomalous, and Magnetospheric Particle Explorer (SAMPEX). Launched in 1992, SAMPEX has returned a wealth of data on the low-altitude radiation environment. The satellite is in a high inclination (82°) polar orbit with an altitude of 520 × 670 km. Its orbit passes through many L-shells, and its data, although not from a high altitude, contain information from those L-shells. The SAMPEX Proton/Electron Telescope (PET) provides measurements on precipitating electrons from 0.4 to about 30 MeV over the polar regions. Contact Dr. Dan Baker, phone 303-492-0591.

B.1.2.7 Other Sources. The NASA International Solar-Terrestrial Physics (ISTP) program has several satellites in orbit that are useful for specific orbits (e.g., plasma conditions in the solar wind or in Earth's magnetotail). A Web site is http://www-istp.gsfc.nasa.gov. The European satellite *Giove-A* has a simple but elegant experiment, Merlin, on board that measures electron flux and other plasma parameters. Reference [14] and more recent papers by Ryden and others describe excellent results from this MEO satellite.

For anomaly investigations, it is desirable to determine quickly what the state of the electron environment was during the event. No appropriate plasma data may be available for either that time period or for the particular spacecraft orbit. In that case, possible secondary sources are the geomagnetic indices or anomaly data from other spacecraft in orbit at the same time. These data are also of value as support material in carrying out anomaly investigations, as they may allow identification of the actual cause, such as surface charging or single-event upsets (SEUs). NOAA's World Data Center (WDC) at Boulder, Colorado, provides a number of useful indices on a near real-time basis and maintains a spacecraft anomaly database. These materials can be addressed through the Web at http://www.ngdc.noaa.gov/wdc/.

Interest is increasing in the development of a simple universal space environment detector for flight on commercial spacecraft to monitor surface and internal charging fluxes. The International Telecommunications Satellite Organization (INTELSAT) has flown at least one such device; others have been flown as well. If a net of such sensors should become available, it might be possible to provide real-time measurements of the state of Earth's plasma and radiation environments and forecast surface and internal discharging effects.

B.2 GEOSYNCHRONOUS ENVIRONMENTS

B.2.1 Geosynchronous Plasma Environments

In this section, the geosynchronous plasma environment is described in terms of temperature and number density. This simple characterization of the environment assumes two species, electrons and protons, where the energy distribution of each species is described by the Maxwell–Boltzmann distribution (Section B.1.1). This treatment is used because the Maxwell–Boltzmann function can easily be used to calculate spacecraft charging. If the Maxwell–Boltzmann distribution is not used, actual data should be curve-fit digitally and integrated numerically at a much greater computational cost. If a single Maxwell–Boltzmann distribution is inadequate for a given circumstance, the data measured are often treated as the sum of two Maxwell–Boltzmann populations. Species such as oxygen and helium can be included as additional Maxwellian populations. (*Note:* Other representations, such as a kappa distribution, are also possible, but the Maxwell–Boltzmann distribution is adequate for most simple charging estimates.)

Next, we describe in greater detail the characterization of the geosynchronous plasma environment in terms of Maxwell–Boltzmann distribution and its moments. The interested reader is also referred to more recent studies of the charging environment using data from the LANL electron and ion spectrometers on a number of geosynchronous spacecraft. See, for example, references [13] and [12] for the ~1-eV to ~45-keV electron and ion environments and reference [10] for the corresponding 30-keV to 2.5-MeV electron environment (the POLE model). Reference [11] has merged the LANL data with data from the Japanese Data Relay Test Satellite to cover the range from 1 keV to 5.2 MeV (the IGE-2006 model).

An initial step in characterizing environments is to consider averages. Ten-minure averages of approximately 45 days per spacecraft were estimated from the ATS-5, ATS-6, and SCATHA (experiment SC9) spacecraft. The corresponding averages (Table B-1) and standard deviations (Table B-2) for each spacecraft were then estimated. The ions were assumed to be protons in these tables. Note that in many cases, the standard deviation exceeded the average. This resulted from the great variability of the geosynchronous environment and illustrates the inherent difficulty of attempting to characterize the "average" plasma environment. (Another way of characterizing the data that avoids some of these problems is to assume that the data are statistically log-normally distributed.) These values are useful, however, in estimating the mean or pre-storm conditions that a spacecraft will experience, as the initial charge state of a spacecraft is important in determining how the vehicle will respond to a significant environmental change. Also, these averages give an approximate idea of how plasma conditions vary

Table B-1. Average Parameters from Selected Spacecraft

Parameter	ATS-5	ATS-6	SCATHA
	Electron Parameters		
Number density (cm^{-3})	0.80	1.06	1.09
Current density (nA/cm^2)	0.068	0.096	0.115
Energy density (eV/cm^3)	1970	3590	3710
Energy flux ($eV/cm^2 \cdot s \cdot Sr$)	0.98×10^{12}	2.17×10^{12}	1.99×10^{12}
Number density for population 1 (cm^{-3})	0.578	0.751	0.780
Temperature for population 1 (keV)	0.277	0.460	0.550
Number density for population 2 (cm^{-3})	0.215	0.273	0.310
Temperature for population 2 (keV)	7.04	9.67	8.68
Average temperature (keV)	1.85	2.55	2.49
Root-mean-square temperature (keV)	3.85	6.25	4.83
	Ion Parameters (Assumed to Be Primarily H^+)		
Number density (cm^{-3})	1.36	1.26	0.58
Current density (pA/cm^2)	5.1	3.4	3.3
Energy density (eV/cm^3)	13,000	12,000	9,440
Energy flux ($eV/cm^2 \cdot s \cdot Sr$)	2.6×10^{11}	3.4×10^{11}	2.0×10^{11}
Number density for population 1 (cm^{-3})	0.75	0.93	0.19
Temperature for population 1 (keV)	0.30	0.27	0.80
Number density for population 2 (cm^{-3})	0.61	0.33	0.39
Temperature for population 2 (keV)	14.0	25.0	15.8
Average temperature (keV)	6.8	6.3	11.2
Root-mean-square temperature (keV)	12.0	23.0	14.5

over a solar cycle since the ATS-5 data are for 1969-70, the ATS-6 data for 1974-76, and the SCATHA data for 1978.

A second way of considering environments is to look at worst-case situations. In addition to Table B-1, several worst-case estimates of the parameters have been made for the geosynchronous environment (Table I.1). These values were derived from fits to actual plasma distributions observed during the several known worst-case ATS-6 and SCATHA charging events. The SCATHA spacecraft instrumentation allowed a breakout of the data into components parallel and perpendicular to the magnetic field and thus permitted a more realistic representation of the actual environment. These values are particularly useful in estimating the extremes in environment that a geosynchronous spacecraft is likely to encounter and are described in Appendix I.

A third quantity of interest in estimating the effects of the space environment on charging is the yearly percentage of occurrence of the plasma parameters. The occurrence frequencies of the temperature and current (Fig. B-1) were derived by fitting the observed distributions of electron and ion temperature for UCSD instruments on ATS-5, ATS-6, and SCATHA. The figures are useful in estimating the time during the year that a specified environment might be expected.

The fourth and a very important quantity of interest is how the plasma parameters vary with time during a charging event. The approaches determining this

Table B-2. Standard Deviations

Parameter Standard Deviation (±)	ATS-5	ATS-6	SCATHA
Electron Standard Deviations			
Number density (cm^{-3})	0.79	1.1	0.89
Current density (nA/cm^2)	0.088	0.09	0.10
Energy density (eV/cm^3)	3,100	3,700	3,400
Energy flux ($eV/cm^2 \cdot s \cdot Sr$)	1.7×10^{12}	2.6×10^{12}	2.0×10^{12}
Number density for population 1 (cm^{-3})	0.55	0.82	0.70
Temperature for population 1 (keV)	0.17	0.85	0.32
Number density for population 2 (cm^{-3})	0.38	0.34	0.37
Temperature for population 2 (keV)	2.1	3.6	4.0
Average temperature (keV)	2.0	2.0	1.5
Root-mean-square temperature (keV)	3.3	3.5	2.9
Ion Standard Deviations (Assumed to Be Primarily H+)			
Number density (cm^{-3})	0.69	1.7	0.35
Current density (pA/cm^2)	2.7	1.8	2.1
Energy density (eV/cm^3)	9,700	9,100	6,820
Energy flux ($eV/cm^2 \cdot s \cdot Sr$)	3.5×10^{11}	3.6×10^{11}	1.7×10^{11}
Number density for population 1 (cm^{-3})	0.54	1.78	0.16
Temperature for population 1 (keV)	0.30	0.88	1.0
Number density for population 2 (cm^{-3})	0.33	0.16	0.26
Temperature for population 2 (keV)	5.0	8.5	5.0
Average temperature (keV)	3.6	8.4	4.6
Root-mean-square temperature (keV)	4.8	8.9	5.3

quantity range from detailed models simulating the magnetosphere to averages over many geomagnetic storms. For design purposes, we have adopted a simulation of the electron and proton current and temperature that approximates the natural variations in the potential as predicted by charging analysis codes. A time-history sequence suitable for modeling the worst effects of a geomagnetic storm is presented in Fig. B-2.

B.2.2 Geosynchronous High-Energy Environments

Unlike the plasma environment, the high-energy electron geosynchronous environment (GEO) is perhaps the best characterized of Earth orbits because of its importance for communications satellites. Quantitative data for GEO are more readily available than for other orbits. There are, however, a number of characteristics of the environment that need to be considered. These range from variations with longitude to rapid time-dependent variations in the high-energy electron spectra. Each of these is discussed below.

B.2.2.1 Variation with Solar Cycle. The high-energy electron population at GEO has a long-term variation with the solar or, more commonly, the sunspot

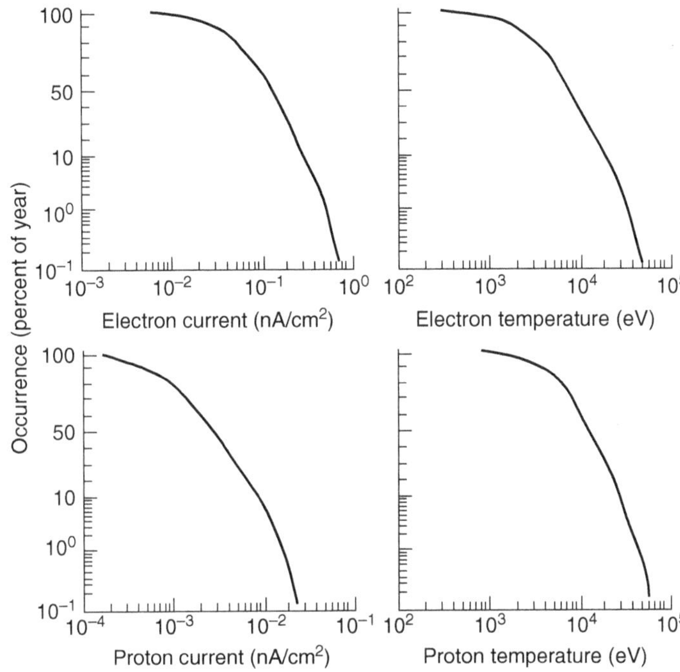

Fig. B-1. Occurrence frequencies of geosynchronous plasma parameters. (From [15].)

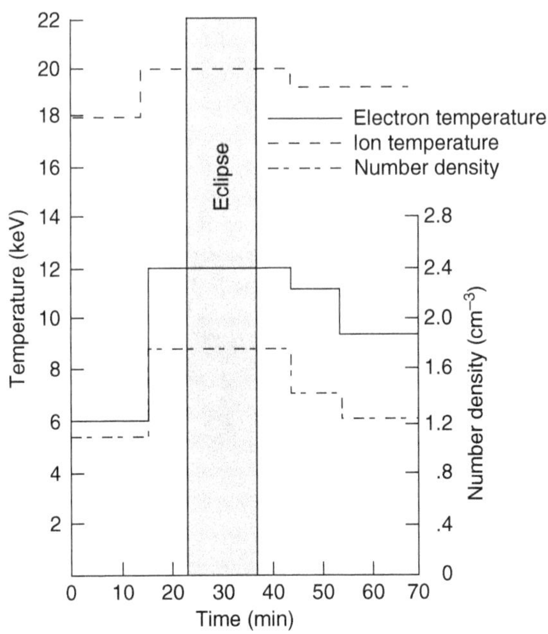

Fig. B-2. Suggested time history for simulating a substorm. (From [15].)

cycle (about 11 years). The $E > 2$ MeV electron population as measured by the geosynchronous GOES-7 satellites is roughly anticorrelated with the sunspot cycle; when the solar sunspot number is low, the GOES $E > 2$ electron flux is high. This is shown in Figs. B-3 and B-4.

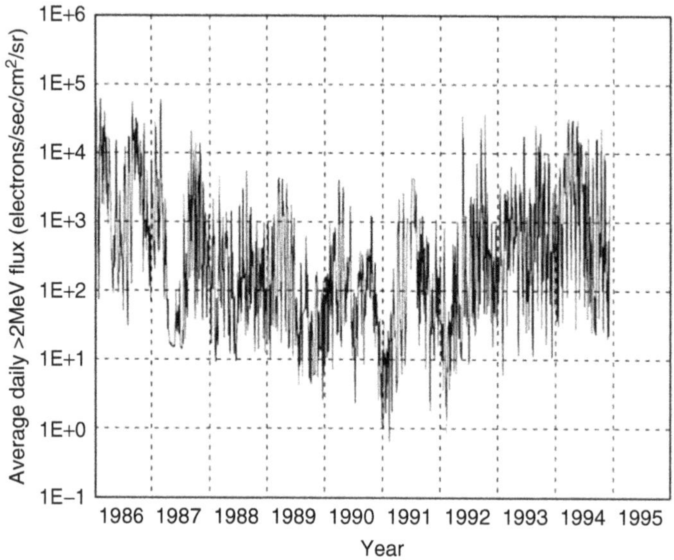

Fig. B-3. Average flux at geosynchronous orbit for $E > 2$ MeV electrons as measured by the GOES spacecraft over approximately one solar cycle (1986–1995). (From [16].)

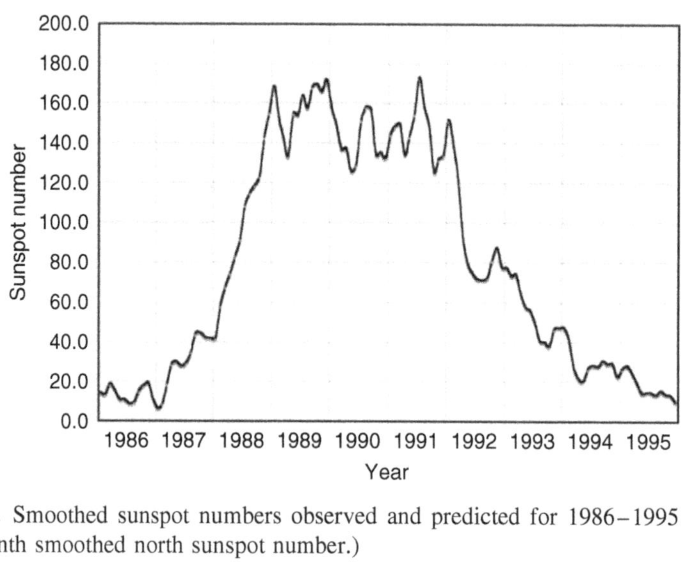

Fig. B-4. Smoothed sunspot numbers observed and predicted for 1986–1995 (monthly three-month smoothed north sunspot number.)

Flying a mission at solar maximum would imply a lower mission (>2 MeV) fluence/dose. Unfortunately, most GEO missions nowadays have durations much longer than five years; therefore, for projects with an unknown launch date, the satellite should be designed to withstand the worst of these periods. This can be a problem, however, as the range between the worst-case conditions and the least stressing is more than 100 : 1 in energetic electron flux. However, the Sun, which drives these environments, does not strictly obey averages, and even during times when the >2 MeV electron fluxes are usually low, the energetic electron fluxes can be extremely high. The project manager, knowing the mission schedule, may wish to assume some risk to save project resources, but the authors advise against such a strategy.

B.2.2.2 Variation with Longitude. The plasma/radiation environment is linked to Earth's magnetic field lines. Magnetic field lines are described in terms of *L*-value, the distance that a given magnetic field line crosses the magnetic equator in Earth radii (referenced to a dipole magnetic field model). Following a particular field line as it rotates around Earth traces out a surface called an *L-shell*. As charged particles (electrons, protons, etc.) are trapped to first order on a magnetic field line/L-shell, the radiation flux can be described in terms of the magnetic field strength at the observation point and the L-shell that passes through the point; this B–L coordinate system is often used in modeling radiation belts. Because Earth's magnetic dipole is tilted and offset with respect to the Earth's rotational axis, real Earth B–L values vary in longitude around geosynchronous orbit (Fig. B-5). Because the radiation environment is approximately constant on a particular L-shell at the magnetic equator, there is a change in the radiation environment at different longitudes as different B–L values are encountered at GEO altitudes. The corresponding fluence and dose variations at GEO are shown in Fig. B-6.

The GEO electron fluences in Fig. B-6 are for the AE8 model, while the dose from electrons is for the CRRESRAD model. This figure is shown only to illustrate the average longitudinal variation. The maximum electron environment should be used for all satellites, even if their longitudinal location is known.

B.2.2.3 Variation with Averaging Interval. In addition to long-term solar cycle variations, there are short-term temporal variations associated with geomagnetic activity and rapid changes in Earth's magnetosphere. As a consequence, the average high-energy electron flux varies with the time interval over which the averaging is carried out. This can be seen when a large data set, gathered with a high time resolution, is averaged over increasingly longer integration times. The GOES $E > 2$ MeV electrons are returned with a 5-min resolution. The variation between the daily peak flux determined in a 5-min interval to the peak flux average in a 24-h period is about 3 to 4 (the 24-h average peak is, as would be expected, lower). This issue of averaging interval should be kept in mind when comparing different data sets. Analysis of Fig. B-3 data from Herbert Sauer gives a similar answer (Fig. B-7).

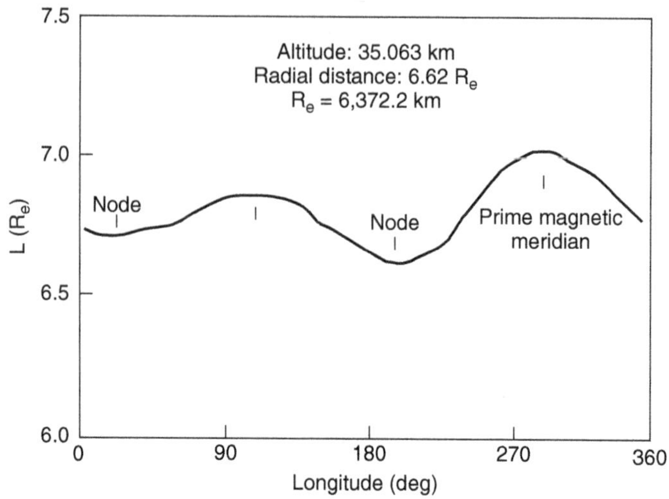

Fig. B-5. L-shell values (units of Earth radii) around Earth's equator (0° latitude) versus east longitude. (From [17].)

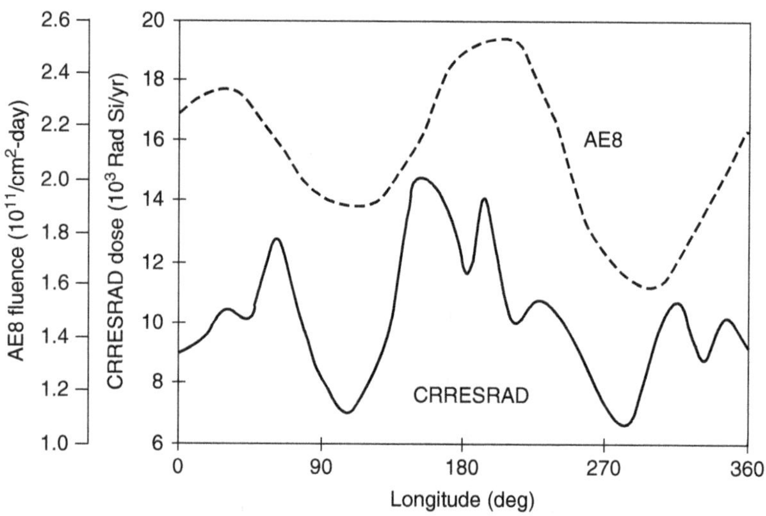

Fig. B-6. AE8 >0.5 MeV daily electron fluence and CRRESRAD annual dose caused by > 1-MeV electrons plotted as functions of satellite east longitude at $6.6R_e$ for the AE8 (>0.5 MeV) and CRRESRAD (>1 MeV) models. (From [18].)

B.2.2.4 Variation with Local Time. The high-energy electrons at a given geosynchronous longitude vary daily with local time. On active days, the flux variation is about 10 : 1 from local noon to local midnight, with the highest flux near local noon. (The NOAA Web site, http://www.sec.noaa.gov/today.html,

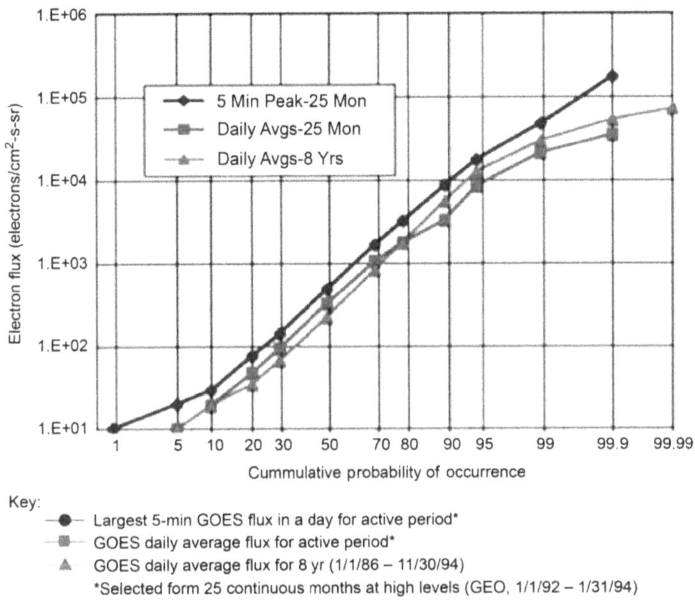

Fig. B-7. Cumulative probability of occurrence of GOES-7 $E > 2$ MeV electron fluxes for several different assumptions. (*See insert for color representation of the figure.*)

shows the current 5-min electron flux at GEO for the last 3-day interval). The normal 24-h average of the GOES

$E > 2$ MeV electron flux (e/cm$^2 \cdot$ s $\cdot$ Sr) is about one-third of the peak daily flux (the highest flux in a 5-min period) in these plots.

B.2.2.5 Spectrum Variation. The integral electron spectrum varies with time in both shape and amplitude. Figure 2-6 presents a worst-case high-amplitude energy spectrum from the LANL SOPA detectors averaged over a few hours compared with a spectrum predicted by the AE8 model, which is a long-term average. Data from the AE8 average show a different spectral shape as well as lower amplitudes. That is, the ratio of integral electron flux at 2 MeV to that at 600 keV is generally not the same from day to day. It can be seen that whereas at low energies ($E < 100$ keV), the curves approach each other, above 1 MeV the spectra rapidly diverge, with the worst-case spectrum approximately two orders of magnitude higher than the AE8 spectrum. This large difference between nominal, time-averaged, and short-term worst-case conditions is characteristic of the radiation environment at Earth. Because of its long-term averaging interval (about five years), the AE8 model is inappropriate for internal charging calculations, as the effects typically are on the order of days or less. The effects of radiation-induced conductivity have not been included in the statements above. Radiation-induced conductivity will reduce the internal electric field. The effect may become noticeable at about 2 MeV, but not enough material data are available to make use of that fact.

B.2.2.6 Amplitude Statistics. An excellent set of data for the statistical analysis of the long-term variations in the total electron flux at geosynchronous orbit is that from the NOAA GOES-7. The data are only available for electrons for $E > 2$ MeV, but the measurements are from one detector and available for approximately one complete solar cycle (Fig. B-3). Figure B-7 plots the cumulative probability of occurrence of GOES-7 electron fluxes. The time span was an eight-year period encompassing the largest energetic fluxes in that solar cycle. Figure B-7 shows amplitude statistics for three statistics from that data set as follows:

1. For the worst 25 months, the day's highest 5-min average flux.
2. For the worst 25 months, the daily average flux.
3. For the entire eight years, the daily average flux.

The circles are the peak GOES electron flux data (largest-amplitude 5-min value in the day) for times of higher flux (January 1, 1992, through January 31, 1994). The triangles correspond to the cumulative probability for the daily GOES average fluxes over the eight-year span from 1986 to 1994. The squares correspond to the GOES data for all daily averages from January 1, 1992, through January 31, 1994. All data are from reference [16]. The key feature to be noted here is that a Gaussian probability distribution implied by a straight-line fit from about 10% to about 95% does not explain the data above the 95th percentile. This makes it difficult to extrapolate with any confidence to a 99.99th percentile environment. The fall-off at the higher percentiles is real [19]. Thus, the worst environments, although real, are less frequent than a simple Gaussian distribution would imply. The reader is cautioned about trying to use these probabilities for design purposes; use the worst-case energy spectrum of Fig. 2-6.

B.3 OTHER EARTH ENVIRONMENTS

B.3.1 MEO

Medium Earth orbit (MEO) ranges from roughly 2000 to 25,000 km (1240 to 15,500 mi) altitude, with an electron flux peak at about 20,000 km (12,400 mi) altitude (the inner electron belt). For internal charging it is the most stressing of the Earth environments. As the Global Positioning Satellite (GPS), as well as some of the multispacecraft communications systems proposed, fly in this orbit, it is a major environment of concern in the study of internal charging phenomena. Figure B-8 is a meridional schematic of Earth's radiation belts at 0° longitude that shows the AE8 and AP8 predictions of the electron ($E > 1$ MeV) and proton ($E > 10$ MeV) fluxes. This plot clearly shows the two-belt structure of the electron belts and the horns that extend down to lower altitudes (the poles). It gives a clear picture of the MEO environment and how it is related to orbital characteristics. Each region has a unique spectrum associated with it, which would affect internal charging calculations. It should also be noted that a third electron belt can sometimes appear between the two main belts after severe geomagnetic storms. This belt can last for months before disappearing.

Note: Figure B-8 shows both electron and proton fluxes, as referenced to Earth's idealized dipole magnetic coordinates, combined onto one chart. The vertical axis is the pole axis with vertical units of Earth radii. The horizontal scale is magnetic equatorial distance from the axis in Earth radii. The upper half-chart represents protons; the southern hemisphere proton flux is a mirror image. The electrons (lower half-chart) also are symmetric above and below the magnetic equator in this coordinate system.

B.3.2 PEO

A second important orbital regime is that associated with highly inclined polar orbits. As seen in Fig. B-8, a polar orbit at low altitudes can pass through the horns of the electron belts and experience a significant, if short-duration, flux of high-energy electrons. Many military spacecraft, most imaging spacecraft, and low-altitude communications fleets are in polar orbits. For low-altitude orbits [< 1000 km (620 mi)], the risk of internal charging is present but generally much lower than at GEO or MEO. At higher altitudes, the interaction is dependent on the

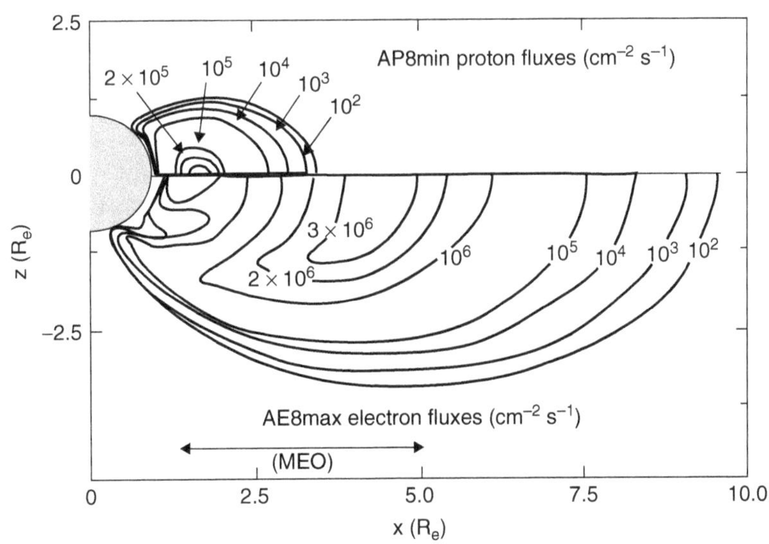

Fig. B-8. Schematic of Earth's radiation belts as estimated by the AE8 and AP8 models; contours for $E > 1$ MeV electrons and $E > 10$ MeV protons for 0° longitude. (Adapted from [20].)

details of the orbit and can be minimized with a proper choice of eccentricity and inclination. Even so, any high-inclination orbit should be evaluated for potential internal charging issues early in the mission design.

B.3.3 Molniya Orbit

Another common orbit for Russian spacecraft is the Molniya orbit. A Molniya orbit follows an elliptical track with a perigee of 500 km (310 mi) and an apogee of 39,000 km (24,000 mi). This orbit is inclined at 63° and the period is on the order of 12 h. As a spacecraft spends most of its time at apogee, this orbit provides good ground coverage for long periods of time at high latitudes (e.g., over Russia). In this orbit, satellites traverse a full range of space environments: from the higher-density, low-energy plasma at LEO through the radiation belts to interplanetary environments. The orbit is also exposed to light and dark so that the satellite is subjected to all environmental variations. Again, the high-energy electron environment should be evaluated for possible internal charging issues for Molniya missions.

B.4 OTHER SPACE ENVIRONMENTS

B.4.1 Solar Wind

Aside from the energetic particle doses from sporadic solar proton events (SPEs), which are not particularly relevant to either surface or internal charging, the solar wind environment is relatively benign for most spacecraft charging applications. The solar wind is a fully ionized, electrically neutral, magnetized plasma that flows outward from the Sun. Table B-3 summarizes many of the characteristics of the solar wind in the ecliptic plane. Perhaps not clear from the table is that the

Table B-3. Characteristics of the Solar Wind at 1 AU in the Ecliptic Plane

Property	Minimum	Maximum	Average
Flux (n/cm$^2 \cdot$ s)	10^8	10^{10}	2 to 3 $\times 10^8$
Velocity (km/s)	200	2500	400 to 500
Density (n/cm^3)	0.4	80	5 to $>$ 10
Temperature (eV)	0.5	100	20
T_{max}/T_{av}	1.0 (isotropic)	2.5	1.4
Helium ratio (N_{He}/N_H)	0	0.25	0.05
Flow direction	±15° from radial		~ 2°E
Alfven speed (km/s)	30	150	60
B (nT)	0.25	40	6
B vector	Polar component		Average in ecliptic plane
	Planar component		Average in spiral angle ~ 45°

Source: [21].

solar wind is highly variable and is coupled to the 11-year solar cycle of activity. Recent years have seen the creation of an interplanetary system of solar wind weather stations designed to closely monitor both solar and solar wind activity (e.g., Ulysses, WIND, Solar and Heliospheric Observatory (SOHO), Yohkoh Observatory, Advanced Composition Explorer (ACE), and the Transition Region and Coronal Explorer (TRACE)). One of these, *Ulysses*, has flown over the poles of the Sun and mapped the solar wind in three dimensions. These spacecraft have identified a variety of characteristic features associated with the solar wind plasma. Of particular interest are the coronal mass ejection (CME) events and the high-speed solar wind streams, as these tend to dominate what might be termed extreme conditions. These are illustrated in Fig. B-9 and demonstrate the variability of the solar wind. It has, indeed, proven difficult, if not impossible, to define one or two worst-case solar wind charging environments, given the rich variety of plasma conditions and the potentially unique charging response of any given spacecraft design to those environments.

Minow, Parker, and their colleagues have carried out an in-depth review of the *Ulysses* and similar data solar wind data. They have generated reference spectra for the solar wind electron and proton environments from the *Ulysses* data in terms of frequency of occurrence percentiles (Fig. B-10). These spectra can be used to estimate surface and internal charging in the solar wind. As this level of detail is not needed in general for the surface charging studies, Maxwell–Boltzmann distributions can be assumed instead. Representative solar wind parameters under this assumption are tabulated for 1 AU and 0.5 AU in Table B-4. (*Note:* For simplicity, only the core population for the solar wind electrons was considered, while the electron halo population was ignored.) Nominal solar wind properties for these two environments are listed in Table B-4.

B.4.2 Earth, Jupiter, and Saturn Magnetospheres Compared

Table B-5 lists the principal characteristics of the terrestrial, Jovian, and Saturnian magnetospheres. Jupiter and Saturn are roughly 10 times the size of Earth, while their magnetic moments are, respectively, 2×10^4 times and 500 times larger. As the magnetic field at the Equator is proportional to the magnetic moment divided by the cube of the radial distance, the terrestrial and Saturnian magnetospheres scale similarly in terms of planetary radii. The Jovian magnetic field, however, is 20 times proportionally larger. An additional consideration is that the photoelectron flux at 1 AU for the Earth is about 25 times that at Jupiter ($\sim$5 AU) and about 100 times that at Saturn ($\sim$10 AU).

The rotation rate is also an important factor. Both Jupiter and Saturn spin over twice as fast as Earth ($\sim$10 h versus 24 h). Given their strong magnetic fields, this means that the cold plasma trapped in these magnetospheres is forced to co-rotate at velocities much higher than a spacecraft's orbital velocity. This is opposite to the situation at Earth, where at low altitudes a spacecraft orbits about 8 km/s faster than the ionospheric plasma. Co-rotation velocities can range from 30 to 40 km/s near Jupiter and Saturn to over 100 km/s in their outer

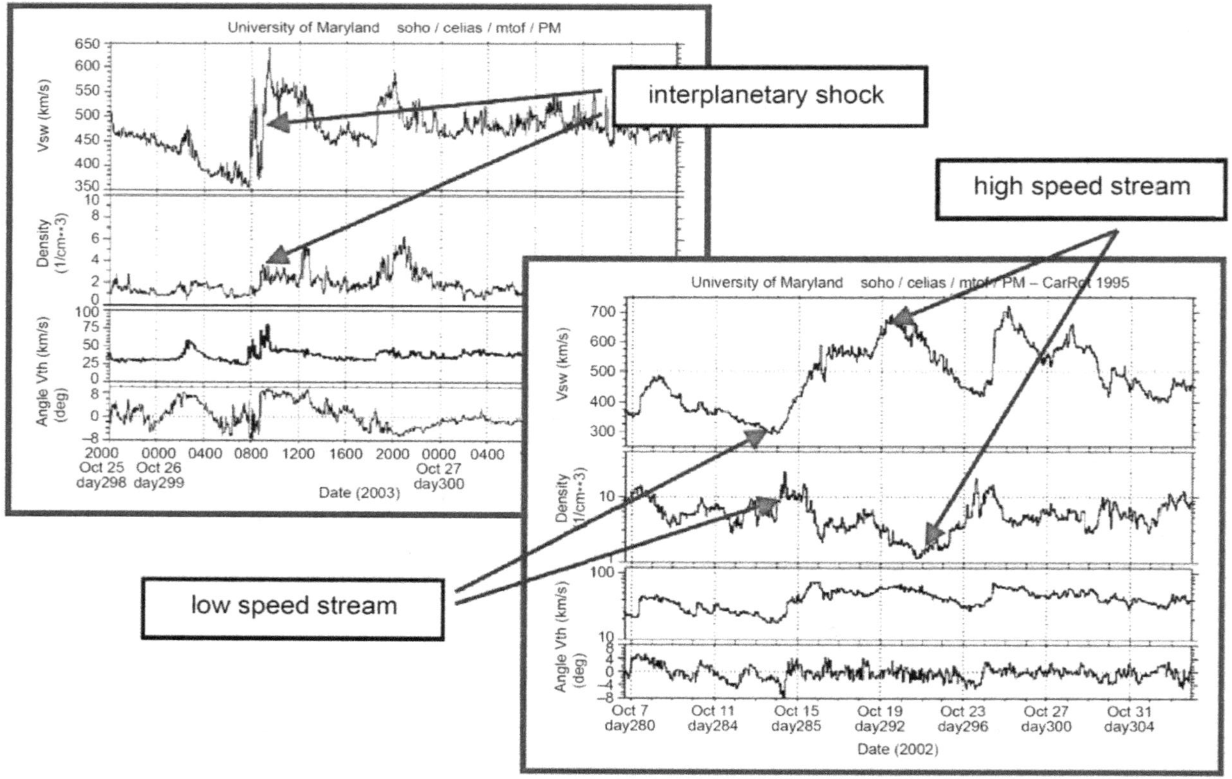

Fig. B-9. Solar wind parameters for an interplanetary shock and high- and low-speed stream versus time as measured by the SOHO spacecraft. Based on the CELIAS/MTOF experiment on the Solar Heliospheric Observatory (SOHO) spacecraft. SOHO is a joint mission of the European Space Agency and the National Aeronautics and Space Administration. (From [22]; SOHO CELIAS/Proton Monitor data, courtesy of University of Maryland).

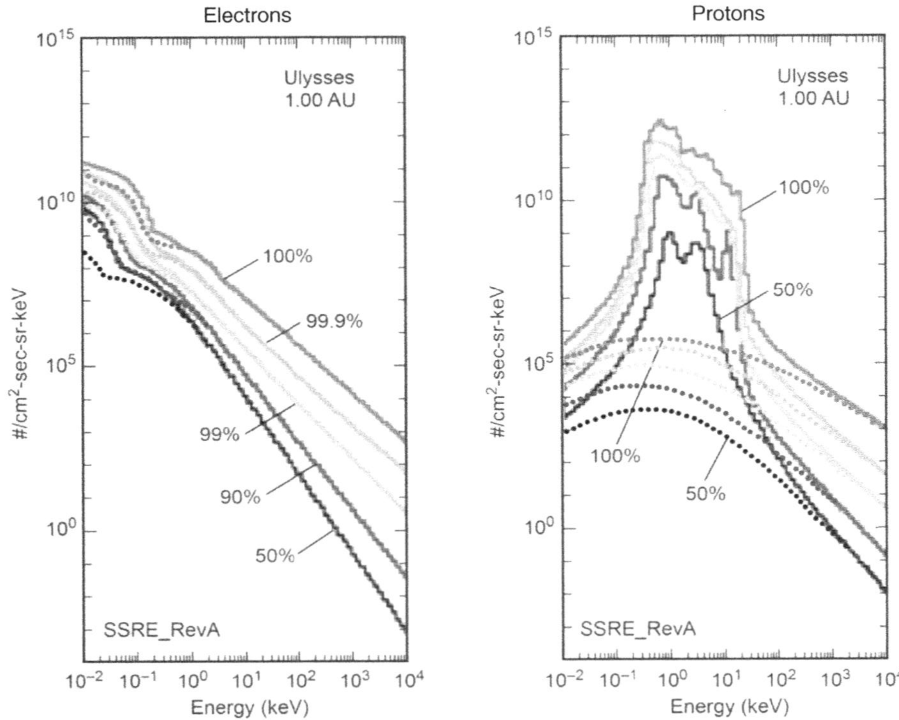

Fig. B-10. Solar wind particle spectra based on measurements made by the *Ulysses* spacecraft for environments of various probabilities. The solid lines represent flows from the Sun; dashed lines represent flows toward the Sun. (From [23].) (*See insert for color representation of the figure*.)

Table B-4. Nominal Solar Wind Plasma Environments

Plasma Environment[a]	0.5 AU	1.0 AU
RE (cm^{-3})	17	12.8
TE (eV)	10.6	11.13
RI (cm^{-3})	17	12.8
TI (eV)	40	10
Photoelectron current (CPH) (nA/cm^2)	8	2
Bulk flow velocity (km/s)	702	327
Potentials (estimated):	**0.5 AU**	**1.0 AU**
Shadowed (insulator)	−22	−22.6
Sunlight (conductive)	11.7	7.5

[a]RE, density for electron plasma population; TE, temperature for electron plasma population; RI, density for ion plasma population; TI, temperature for ion plasma population.

Table B-5. Magnetospheres of Earth, Jupiter, and Saturn

	Region/Parameter			
Planet	Equatorial Radius (km; mi)	Magnetic Moment (G-cm^3)	Rotation Period (h)	Aphelion/Perihelion (AU)
Earth	6.38×10^3; 3960	8.10×10^{25}	24.0	1.01/0.98
Jupiter	7.14×10^4; 44,400	1.59×10^{30}	9.925	5.45/4.95
Saturn	6.00×10^4; 37,000	4.30×10^{28}	10.23	10.06/9.01

magnetospheres. As the magnetosphere is the primary controlling factor for the local plasma environments, the charging environment differs considerably for each of these planets.

The magnetosphere of Jupiter is dominated by the following three factors:

1. The magnetic field tilt (11°) relative to its spin axis
2. Its rapid rotation
3. The Jovian moon Io at $5R_j$

Io generates a vast torus of gas. The rapid rotation of Jupiter's magnetic field forces the cold plasma associated with this torus to accelerate and expand by centrifugal force into a giant disk. The magnetic field tilt and rotation rate make this plasma disk move up and down, so at a given location, plasma parameters vary radically over a 10-h period (or 5 h in the plasma sheet). Jupiter's environment can be roughly divided into the following three populations:

1. The cold plasma associated with the Io torus and the plasma disk ($0 < E < 1$ keV)
2. The intermediate plasma and aurora (1 keV $< E <$ 100 keV)
3. The radiation environment ($E >$ 100 keV)

The cold plasma environment has high densities ($\sim$2000 cm^{-3}) and low energies (1 eV to 1 keV). This plasma consists of hydrogen, oxygen (singly and doubly ionized), sulfur (singly, doubly, and triply ionized), and sodium (singly ionized) ions. The intermediate plasma environment is made up of electrons ($\sim$1 keV) and protons ($\sim$30 keV) and assumed to vary exponentially from $\sim$5 cm^{-3} for $r < 10R_j$ to 0.001 cm^{-3} beyond $40R_j$. Co-rotation velocities vary from about 45 km/s at $4R_j$ to about 250 km/s at $20R_j$.

Saturn is marked by a magnificent set of rings that are its most obvious feature and set it apart from all the other planets. Aside from the rings, however, Saturn's magnetosphere resembles Jupiter's—a cold inner plasma disk giving way to a lower-density, slightly higher-energy plasma disk at large distances. Although there is no Io-equivalent moon in the inner magnetosphere, there is still a fairly dense cold plasma sheet and, at $\sim$20 R_s, Saturn's huge moon Titan contributes a large cloud of neutral gas in the outer magnetosphere. Unlike Jupiter, Saturn's magnetic field axis is apparently aligned with the spin axis so that the plasma ring around Saturn is relatively stable compared to that of Jupiter. Plasma co-rotation velocities are similar to those of Jupiter, although maximum velocities tend to peak a little above 100 km/s.

A simple design tool based on current balance and on Earth's, Jupiter's, and Saturn's cold and intermediate plasma environments (the latter also includes the aurora that have been observed at all three planets) has been used to estimate the spacecraft-to-space potentials for these planets. The results of this tool for a spherical spacecraft with aluminum surfaces are presented in Table B-6 for

Table B-6. Representative Charging Levels at Earth, Jupiter, and Saturn Based on a Simple Charging Design Tool

Region	Plasma Convection Velocity (km/s)	Potential (in Sunlight) (v)	Potential (No Sun/No Secondaries) (v)
Earth			
Ionosphere	8	−0.7	−4.4
Plasmasphere	3.7	−1.6	−3.8
Auroral zone	8	−0.7	−500
Geosynchronous	3	2.0	−20, 000
Jupiter			
Cold torus	44	−0.59	−1.2
Hot torus	100	−60	−70
Plasma sheet	150	−94	−130
Outer magnetosphere	250	9.5	−2, 500
Saturn			
Inner plasma sheet	40	~5	−30
Outer plasma sheet	80	~5	−500
Hot outer magnetosphere	100	−100	−500

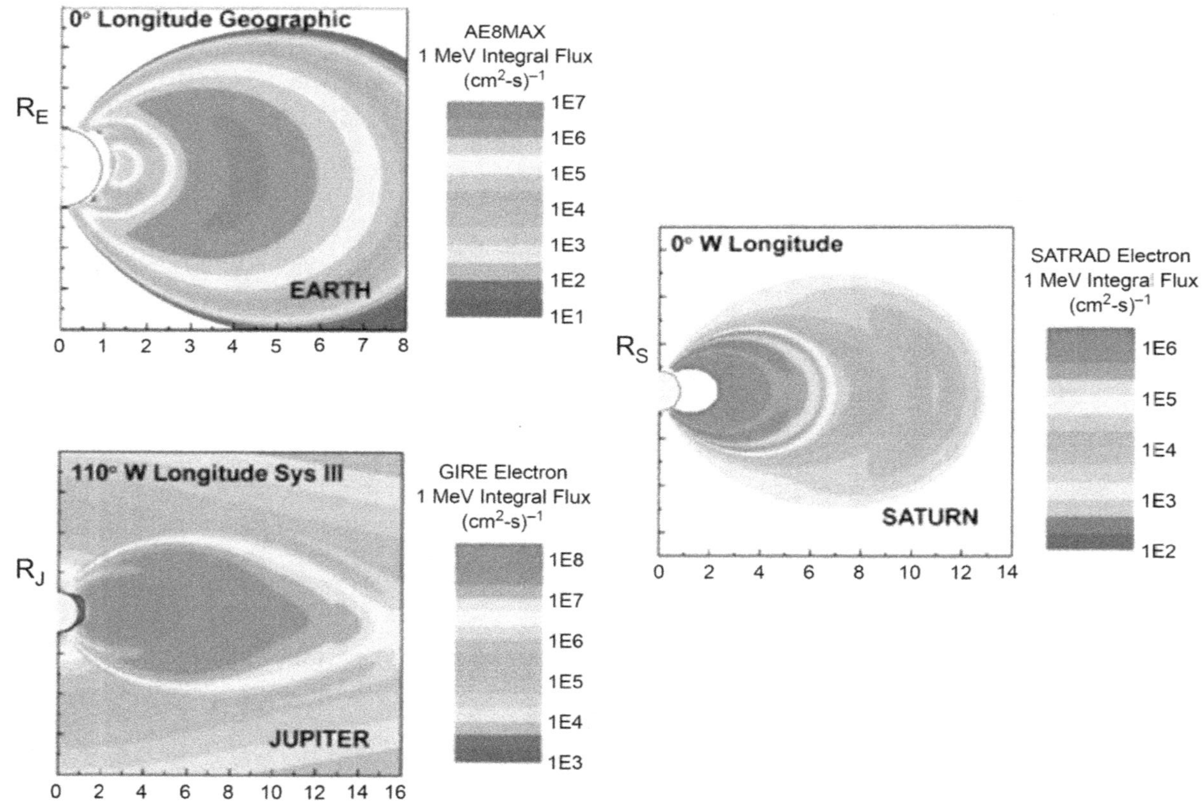

Fig. B-11. 1-MeV electron omnidirectional flux contours for Earth, Jupiter, and Saturn (models used: AE8Max, GIRE, and SATRAD). (Courtesy of Insoo Jun, personal communication.) (*See insert for color representation of the figure.*)

several different plasma regions and situations. Based on this table, Earth clearly represents the worst threat to spacecraft. Negative potentials as high as 20,000 V are predicted near geosynchronous orbit in eclipse, and indeed, potentials in excess of $-20,000$ V have apparently been observed. At Jupiter, potentials are more moderate. Large potentials are observed only if secondary emissions can be suppressed, unlikely but possible for some surface configurations. Conditions at Saturn are similar to those at Jupiter, although somewhat lower in general. Even so, spacecraft surface charging is still a concern for spacecraft survivability at these planets. Indeed, as potentials of even a few tens of volts can seriously affect low-energy plasma measurements, spacecraft charging should be considered for scientific missions to these planets.

The high-energy electrons that are part of the radiation environment at each of the three planets are the source of internal charging. In Fig. B-11, the 1 MeV electron flux contours for Earth (AE8Max model), Jupiter (Galileo Interim Radiation Electron (GIRE) model), and Saturn (Saturn Radiation (SATRAD) model) are presented. In a number of studies [24,25], it has been demonstrated that fluences of 10^{10} e/cm^2 are roughly the level required for an IESD. The fluxes in the most intense regions in Fig. B-11 are on the order of 10^7, 10^8, and 10^6 e/cm$^2 \cdot$ s for Earth, Jupiter, and Saturn, respectively. (*Note:* The inner radiation belt at Saturn is largely missing because of Saturn's ring system.) This implies internal charging times for 1-MeV electrons of about 10^3 s, 10^2 s, and 10^4 s. Flight experience has shown that the Earth poses moderate-to-severe IESD problems, Jupiter has severe IESD, and to date, Saturn has not demonstrated any problems in agreement with these charging times.

REFERENCES

[1] H. B. Garrett and S. E. DeForest, "Analytical Simulation of the Geosynchronous Plasma Environment", *Planetary and Space Science*, vol. 27, pp. 1101–1109, 1979.

[2] H. B. Garrett, D. C. Schwank, and S. E. DeForest, "A Statistical Analysis of the Low-Energy Geosynchronous Plasma Environment: I. Electrons", *Planetary and Space Sciences*, vol. 29, pp. 1021–1044, 1981.

[3] H. B. Garrett, D. C. Schwank, and S. E. DeForest, "A Statistical Analysis of the Low-energy Geosynchronous Plasma Environment: II. Ions", *Planetary and Space Sciences*, vol. 29, pp. 1045–1060, 1981.

[4] H. B. Garrett, D. C. Schwank, P. R. Higbie, and D. N. Baker, "Comparison Between the 30–80keV Electron Channels on ATS-6 and 1976-059A During Conjunction and Application to Spacecraft Charging Prediction", *Journal of Geophysical Research*, vol. 85, no. A3, pp. 1155–1162, 1980.

[5] A. Jursa, Ed., *Handbook of Geophysics and the Space Environment*, 4th ed., Air Force Geophysics Laboratory, Hanscom Air Force Base, Massachusetts, National Technical Information Service Document, Accession AD-A167000, December 5, 1985. Additional reading for the space environment and interactions with spacecraft. An excellent reference for Earth space plasma environments as well as many other space environments.

[6] E. G. Mullen and M. S. Gussenhoven, *SCATHA Environmental Atlas*, AFGL-TR-83-0002, Air Force Geophysics Laboratory, Hanscom Air Force Base, Massachusetts, 1983.

[7] E. G. Mullen, M. S. Gussenhoven, and H. B. Garrett, *A "Worst Case" Spacecraft Environment as Observed by SCATHA on 24 April 1979*, AFGL-TR-81-0231, Air Force Geophysics Laboratory, Hanscom Air Force Base, Massachusetts, 1981.

[8] E. G. Mullen, M. S. Gussenhoven, D. A. Hardy, T. A. Aggson, B. G. Ledly, and E. C. Whipple, "SCATHA Survey of High-Level Spacecraft Charging in Sunlight", *Journal of Geophysical Research*, vol. 91, no. A2, pp. 1474–1490, February 1, 1986.

[9] E. G. Mullen, D. A. Hardy, H. B. Garrett, and E. C. Whipple, "P78-2 SCATHA Environmental Data Atlas", *Spacecraft Charging Technology 1980*, NASA CP 2182/AFGL-TR-81-0270, National Aeronautics and Space Administration, pp. 802–813, 1981.

[10] D. M. Boscher, S. A. Bourdarie, R. H. W. Friedel, and R. D. Belian, "Model for the Geostationary Electron Environment", *IEEE Transactions on Nuclear Science*, vol. 50, no. 6, pp. 2278–2283, 2003.

[11] A. Sicard-Piet, S. Bourdarie, D. Boscher, R. H. W. Friedel, M. Thomsen, T. Goka, H. Matsumoto, and H. Koshiishi, "A New International Geostationary Electron Model: IGE-2006, from 1 keV to 5.2 MeV", *Space Weather*, vol. 6, S07003, doi:10.1029/2007SW000368, 13 pp., 2008.

[12] V. A. Davis, M. J. Mandell, and M. F. Thomsen, "Representation of the Measured Geosynchronous Plasma Environment in Spacecraft Charging Calculations", *Journal of Geophysical Research*, vol. 113, no. A10204, doi:10.1029/2008JA013116, 14 pp., 2008.

[13] M. F. Thomsen, M. H. Denton, B. Lavraud, and M. Bodeau, "Statistics of Plasma Fluxes at Geosynchronous Orbit over More Than a Full Solar Cycle", *Space Weather*, vol. 5, S03004, doi:10.1029/2006SW00025, 9 pp., 2007.

[14] K. A. Ryden, P. A. Morris, A. D. P. Hands, C. S. Dyer, M. Fellows, B. Taylor, C. I. Underwood, D. J. Rodgers, G. Mandorlo, G. Gatti, H. D. R. Evans, and E. J. Daly, "Radiation Monitoring in Medium Earth Orbit over the Solar Minimum Period", Proceedings of *RADECS 2008*, Jyvaskyla, Finland, September 10–13, 2008. Reports and conclusions from data of the Merlin experiment on the *Giove-A* spacecraft (MEO, 23 km, 56° inclination) launched December 28, 2005. Compares to existing space radiation (electrons, dose, protons). Its references are a must-have.

[15] C. K. Purvis, H. B. Garrett, A. C. Whittlesey, and N. J. Stevens, *Design Guidelines for Assessing and Controlling Spacecraft Charging Effects*, NASA Technical Paper 2361, National Aeronautics and Space Administration, September 1984. This document has been widely used by practitioners of this art (usually, EMC engineers or radiation survivability engineers) since its publication in 1984. Its contents are limited to surface charging effects. The contents are valid to this day for that purpose. NASA TP-2361 contents have been incorporated into NASA-STD-4002, Revision A, with heavy editing. Many of the original details, especially time-variant and multiple-case versions of suggested environments, have been simplified into single worst-case environments in NASA-HDBK-4002, Revision A. Some background material has not been transferred into this document, so the original may still be of interest.

[16] *GOES SEM Data Notes: Important Information for Data Users*, website, National Oceanic and Atmospheric Administration National Geophysical Data Center, http://www.ngdc.noaa.gov/stp/satellite/goes/datanotes.html (accessed May 9, 2011).

Herbert Sauer of the National Oceanic and Atmospheric Administration, National Geophysical Data Center (Boulder, Colorado) initially supplied this information to the authors in a personal communication. His data and those of others were later incorporated into the referenced website.

[17] E. G. Stassinopoulos, "The Geostationary Radiation Environment", *Journal of Spacecraft and Rockets*, vol. 17, no. 2, pp. 145–152, March–April 1980.

[18] G. L. Wrenn, "Conclusive Evidence for Internal Dielectric Charging Anomalies on Geosynchronous Communications Spacecraft", *Journal of Spacecraft and Rockets*, vol. 32, no. 3, pp. 514–520, May–June 1995. Note that the author believes that there was still a need to convince people that internal charging was a real phenomenon, as recently as 1995.

[19] C. F. Kennel and H. E. Petschek, "Limit on Stably Trapped Particle Fluxes", *Journal of Geophysical Research*, vol. 71, pp. 1–28, 1966.

[20] E. J. Daly, "The Evaluation of Space Radiation Environments for ESA Projects", *ESA Journal*, vol. 12, pp. 229–247, 1988.

[21] E. N. Parker, *Interplanetary Dynamical Processes*, Interscience, New York, 1963.

[22] H. B. Garrett and J. I. Minow, *Charged Particle Effects on Solar Sails*, NASA Report ISPT-SS-06-101, Marshall Space Flight Center, Huntsville, Alabama, 2007.

[23] J. I. Minow, L. N. Parker, and R. L. Altstatt, "Radiation and Internal Charging Environments for Thin Dielectrics in Interplanetary Space", *9th Spacecraft Charging Technology Conference*, Tsukuba, Japan, April 2–9, 2005.

[24] P. Leung, A. C. Whittlesey, H. B. Garrett, and P. A. Robinson, Jr., "Environment-Induced Electrostatic Discharges as the Cause of *Voyager 1* Power-On Resets", *Journal of Spacecraft and Rockets*, vol. 23, no. 3, May–June, pp. 323–330, 1986. One of the best-documented examples of IESD.

[25] A. R. Frederickson, E. G. Holeman, and E. G. Mullen, "Characteristics of Spontaneous Electrical Discharges of Various Insulators in Space Radiation", *IEEE Transactions on Nuclear Science*, vol. 39, no. 6, pp. 1773–1982, December 1992. This journal paper is a description of the best-known attempt to quantify internal charging effects on orbit by means of a well-thought-out experiment design. The results were not all that the investigators had hoped, but the data are excellent and very good conclusions can be reached from the data, despite the investigators' concerns.

APPENDIX C
Environment, Electron Transport, and Spacecraft Charging Computer Codes

C.1 ENVIRONMENT CODES

Codes are listed below in alphabetical order. Note that some codes do both environments and transport but are listed in one place only.

C.1.1 AE8/AP8

The NASA AE8 (electrons) and AP8 (protons) radiation models are the traditional electron and proton models of Earth's radiation environment. The AE8 predictions for GEO are probably the most used estimates of the average environment. In these codes, the fluxes are long-term averages (about five years or more). There are two versions of each model: AE8 solar minimum and AE8 solar maximum and AP8 solar minimum and AP8 solar maximum. They do not predict the peak electron fluxes that are necessary for the internal charging calculations recommended in this book. Garrett [1] has reviewed the output and problems with the AE8/AP8 models. (*Note:* As of publication of this book, the new AE9/AP9 radiation models had just gone out for beta testing—they were expected to be formally released late in 2011.)

C.1.2 CRRES

CRRES monitored Earth's radiation belts in an eccentric orbit for 14 months starting in July 1990. The data from the spacecraft are in the form of electron and proton flux and dose–depth curves as functions of time and altitude. Environment codes from CRRES include CRRESRAD (dose versus depth); CRRESPRO (proton flux energy spectrum); and CRRESELE (electron flux energy spectrum). They are available from the Air Force Research Laboratory (AFRL). Perform an Internet search on af-GEOSPACE; choose fact sheets: AF-GEOSPACE, and select "software request form" at the bottom of that page.

Guide to Mitigating Spacecraft Charging Effects, First Edition. Henry B. Garrett, Albert C. Whittlesey.

C.1.3 Flux Model for Internal Charging (FLUMIC)

FLUMIC [2], an environmental model developed by ESA and part of the Defense Evaluation and Research Agency (DERA) Internal Charging Threat Analysis Tool (DICTAT), is a position-dependent worst-case model of electron fluxes in the outer radiation belt. The FLUMIC code is explained in the DICTAT user's manual, which can be downloaded from the Web site http://www.spenvis.oma.be/help/models/dictat.html.

C.1.4 GIRE/SATRAD

GIRE and SATRAD environment models are used to estimate the radiation exposure to spacecraft in the out-of-plane radiation environments of Jupiter's and Saturn's magnetospheres, respectively. A time-versus-position trajectory is required as input into the codes. They were developed by NASA/JPL. The source codes and sample inputs/outputs are available from http://www.openchannelfoundation.org/projects/GIRE and /SATRAD.

C.1.5 Handbook of Geophysics and the Space Environment

Jursa's document [3] is an excellent and recommended reference for space environments, including plasma environments for Earth. Even though it was done in 1985, it has not been improved on as a single-source and consistent set of information.

C.1.6 L2 Charged Particle Environment (L2-CPE)

The L2-CPE model is an engineering tool that provides free-field charged-particle environments for the distant magnetotail, magnetosheath, and solar wind environments. L2-CPE is intended for use in assessing contributions from low-energy radiation environments (~0.1 keV to few MeV) to radiation dose in thin materials used in the construction of spacecraft to be placed in orbit about the Sun–Earth L2 point. Reference [4] describes the status of the current version of the L2-CPE model, including the structure of the model used to organize plasma environments into solar wind, magnetosheath, and magnetotail environments, the algorithms used to estimate radiation fluence in sparsely sampled environments, the updated graphical user interface, and output options for flux and fluence environments. Information on the availability of the model can be obtained from J. I. Minow (Joseph.I.Minow@nasa.gov). Other references are [5–7].

C.1.7 MIL-STD-1809, Space Environment for USAF Space Vehicles

Another source of particle estimates is MIL-STD-1809 [8]. This includes electron spectra that can be used in the electron transport codes for estimating IESD. It also has information that supplements Earth environment information in this book.

C.1.8 Geosynchronous Plasma Model

Reference [9] used data from the ATS-5 spacecraft to generate a simple model for analytically simulating the parameters necessary to characterize the geosynchronous plasma. The model is developed in terms of the daily worldwide geomagnetic activity Ap index and local time. Although based on a limited set of ATS-5 data, the simulation adequately models the simultaneous variations in the warm plasma (50 eV to 50 keV) electron and ion populations during injection events. Developed primarily to estimate the varying potentials expected on a shadowed, electrically isolated surface, the simulation can also be employed in a variety of cases where knowledge of the general characteristics of the geosynchronous plasma is necessary. The model has been extended to include data from ATS-6 and the SCATHA spacecraft. Those desiring the latest version of the model should contact Henry Garrett at henry.b.garrett@jpl.nasa.gov (818-354-2644).

C.1.9 Others

Alternative sources of space radiation data include Severn Communications Corporation, 1023 Benfield Boulevard, Millersville, MD 21108 (including AP8 and AE8). Use the Severn Communications Web site to search and find various environmental papers published by their staff. As described in more detail in Section C.2.9, the Space Environment Information System provides an online space environment "handbook" at http://www.spenvis.oma.be/.

C.2 TRANSPORT CODES

Note that some codes do both environments and transport but are listed in one place only.

C.2.1 Cosmic Ray Effects on MicroElectronics 1996 (CREME96)

CREME96 is a Web-based suite of tools hosted at https://creme-mc.isde.vanderbilt.edu/ at Vanderbilt University, Nashville, Tennessee. It incorporates analysis capabilities for the following:

1. Creating numerical models of the ionizing radiation environment in near-Earth orbits:
   a. Galactic cosmic rays (GCRs)
   b. Anomalous cosmic rays (ACRs)
   c. Solar energetic particles (SEPs)
   d. Geomagnetically trapped particles
2. Evaluating the resulting radiation effects on electronic systems in spacecraft and in high-altitude aircraft:
   a. Total ionizing dose (TID)

   b. Displacement damage dose (DDD)
   c. Single-event effects (SEEs), including single-event upsets (SEUs)
3. Estimating the linear energy transfer (LET) radiation environment within manned spacecraft.

The TRANS module of the suite is limited to 1-D and aluminum shielding.

C.2.2 EGS4

EGS4 is a Monte Carlo transport code. The suite is used primarily for electron-beam experiment simulations. It is easy to use and incorporates validated physics models but is limited in geometry modeling and the space environments included. Recent improvements may have added to its capabilities. References [10–12] contain additional information. The Web site for this code is http://www.irs.inms.nrc.ca/EGSnrc/EGSnrc.html at the National Research Council, Institute for National Measurement Standards, Canada.

C.2.3 Geant4

Geant4 is the European counterpart to Monte Carlo N-Particle (MCNP) eXpanded (MCNPX). The Geant family of particle transport codes represents a unique international cooperative effort to model radiation interactions. Many different groups and organizations have contributed specialized analytic components to the basic package. Geant4 is a collection of computer tools for the simulation of the passage of particles through matter. Its areas of application include high-energy, nuclear, and accelerator physics, as well as studies in medical and space science. The two main reference papers for Geant4 are published in *Nuclear Instruments and Methods in Physics Research* [13] and the Institute of Electrical and Electronics Engineers (IEEE) *Transactions on Nuclear Science* [14]. The code and its derivatives make up probably the most sophisticated (and thus complex) modeling package currently available, as it covers a much wider range of problems than space radiation effects. As such, it has a steep learning curve. Special courses and seminars are available for learning its many features. The home page for Geant4 can be found at http://geant4.web.cern.ch/geant4/.

C.2.4 Integrated TIGER Series (ITS)

The ITS code provides electron flux and deposition and has been validated by experiment. It would be the first choice for the electron deposition calculations suggested in this book. Some packages have been simplified to handle simple geometries, such as cylinders and slabs. It apparently has no E-field-induced conductivity parameter. Contact: Radiation Shielding Information Computational Center (RSICC), Oak Ridge National Laboratory, Building 6025, MS 6362, P.O. Box 2008, Oak Ridge, Tennessee (TN) 37831-6362 (ITS CCC-467). One

Web page source is http://rsicc.ornl.gov/codes/ccc/ccc4/ccc-467.html. Another is http://prod.sandia.gov/techlib/access-control.cgi/2004/045172.pdf.

ITS3.0 is a suite of three radiation transport codes that employ a Monte Carlo (mostly forward) technique. The three tools are (1) TIGER (1-D), (2) CYLTRAN (2-D), and (3) ACCEPT (3-D). The codes handle electrons and photons.

C.2.5 MCNP/MCNPE

MCNP is a radiation transport code that employs a Monte Carlo (mostly forward) technique. The code handles neutrons, photons, and electrons. At one institution, the code is used primarily for neutron–photon transport studies. It incorporates a versatile geometry and input/output options. It is, however, slow for space environment applications.

MCNPX is based on MCNP and has the additional capability of handling neutrons, antineutrons, photons, electrons, positrons, muons, antimuons, electron neutrinos, antielectron neutrinos, protons, antiprotons, positive pions, negative pions, neutral pions, positive kaons, negative kaons short, neutral kaons long, deuterons, tritons, helium-3s, and helium-4s (alpha particles). It has been used for proton transport, where secondary particle generation is important.

MCNP/MCNPE, a version of MCNP modified to include transport of electrons, can be used to determine electron flux inside complex spacecraft geometries. MCNP does detailed 3-D Monte Carlo modeling of neutron, photon, and electron transport. MCNPE does 3-D modeling of neutron, photon, and electron transport. They have a powerful geometric capability; however, transport to very deep depths can take extremely long computer runs with a large uncertainty in the results. At shallow depths (as much as 600 mils of aluminum thickness), codes like ITS are preferred. The codes have continual upgrades, so looking at the Web site is advised for the most recent information: http://mcnp-green.lanl.gov/index.html. A new version of this code is MCNP-4B [15].

C.2.6 NOVICE

NOVICE is a charged-particle radiation transport code. It uses an adjoint Monte Carlo technique to model particle fluxes inside a user-specified 3-D shield geometry in particular. NOVICE uses an inside-out particle tracking algorithm. The code handles electrons, photons, protons, and heavy ions ($Z \geq 2$). It can handle fairly complex geometries and is fast as well as easy to use; however, it does not work for secondary particles. Contact: Thomas Jordan, Experimental and Mathematical Physics Consultants, P.O. Box 3191, Gaithersburg, Maryland (MD) 20885, phone 301-869-2317. This source may also have codes for electron deposition calculations [16].

C.2.7 NUMIT

NUMIT, originally developed by A. R. Fredeickson, is a 1-D computer code for estimating internal charging in dielectrics. It computes the full-time dependent current, voltages, and electric fields in the dielectric by iteratively solving a set of equations for monoenergetic photons and electrons normally incident on one side of a dielectric. Contact: Dr. Wousik Kim, Jet Propulsion Laboratory, Mail Stop 122-107, Pasadena, California (CA) 91109.

C.2.8 SHIELDOSE

SHIELDOSE is a charged-particle radiation transport code that calculates the dose inside slab and spherical shield geometries. It also computes dose absorbed in small volumes of some detector materials under specified aluminum shield geometries. See references [2,7]. Web reference: http://modelweb.gsfc.nasa.gov/magnetos/shield.html.

C.2.9 SPENVIS/DICTAT

SPENVIS/DICTAT is a code package is designed for spacecraft internal charging analysis and is available for use on the Web at http://www.spenvis.oma.be/spenvis/ and http://www.spenvis.oma.be/spenvis/help/background/charging/dictat/dictatman.html. DICTAT [18] calculates the electron current that passes through a conductive shield and becomes deposited inside a dielectric. From the charge deposited, the maximum electric field within the dielectric is found. This field is compared with the breakdown field for that dielectric to see if the material is at risk of an ESD.

SPENVIS, a Web-based suite of tools designed for near-Earth analysis, generates either a spacecraft trajectory or a coordinate matrix. In addition to the DICTAT model, it incorporates analysis capabilities for the following:

1. Trapped proton and electron fluxes and solar proton fluences
2. Radiation doses (ionizing and nonionizing)
3. Damage equivalent fluences for Si and GaAs solar panels
4. LET spectra and SEU rates
5. Trapped proton flux anisotropy
6. Atmospheric and ionospheric densities and temperatures
7. Atomic oxygen erosion depths
8. GIRE, the Jovian radiation model

C.2.10 TRIM

TRIM is a radiation transport code that employs a Monte Carlo (forward) technique. It is 1-D and accommodates protons and heavy ions. TRIM is used for proton and heavy-ion beam simulation, and it covers the entire spectrum of heavy-ion types. It is limited to 1-D slab geometry, however, and only incorporates Coulomb interactions.

Table C-1. Properties of the Major Transport Codes

Code Name	Calculates Electron Deposition?	Usable for IESD?	Recommended for IESD Calculation?	Calculates E-Field?	Uses RIC or Conductivity?
DICTAT	—	?	Y	Y	—
EGS4	Y	?	N?	—	—
Geant4	Y	Y	N	N	N
ITS	Y	Y	N	N	N
MCNPX	Y	Y	N	—	—
NUMIT	N	N?	Y	Y	Y

C.2.11 Summary

The preceding transport codes are intended to be used in estimating internal charge deposition—a major step in estimating the probability of IESD. Table C-1 provides a comparison of some IESD charging specific parameters for the major analysis codes. Whereas codes like the TIGER, Geant, and MCNPX allow estimates of the flux (and fluence) with depth in the material, the DICTAT and NUMIT codes estimate the buildup of the fields in the material.

C.3 CHARGING CODES

These codes generally calculate surface charging, potentials, E-fields, and other parameters that are of interest for an overall view of spacecraft charging. Look for one or more that best meets the needs of the project.

C.3.1 Environment Work Bench (EWB)

The EWB code uses simple models of plasma and other space environments and interactions to predict a variety of environmental effects. These include LEO spacecraft floating potentials, as an example. It is international traffic-in-arms regulations (ITAR) restricted. See http://see.msfc.nasa.gov at Marshall Space Flight Center, Huntsville, Alabama (SEE Products: Electromagnetic Effects & Spacecraft Charging).

C.3.2 Multi-Utility Spacecraft Charging Analysis Tool (MUSCAT)

MUSCAT [19] is a Japanese computer code that predicts potentials, with functions similar to those the NASA Charging Analyzer Program (NASCAP).

C.3.3 Nascap-2k and NASCAP Family of Charging Codes

Nascap-2k [20,21] is a widely used interactive toolkit for studying plasma interactions with realistic spacecraft in three dimensions. It can model interactions that occur in tenuous (e.g., GEO orbit or interplanetary missions) and in dense (e.g., LEO orbit and the aurora) plasma environments. Capabilities include surface charging in geosynchronous and interplanetary orbits, sheath and wake structure and current collection in LEO, and auroral charging. External potential structure and particle trajectories are computed using a finite element method on a nested grid structure and may be visualized within the Nascap-2k interface. Space charge can be treated either analytically, self-consistently with particle trajectories, or consistent with imported plume densities. Particle-in-cell (PIC) capabilities are available to study dynamic plasma effects. Material properties of surfaces are included in the surface charging computations. By locating severe surface voltage gradients in a particular design, it is possible to show where discharges could occur. The effect of changes in the surface materials or coatings in those areas on minimizing voltage gradients can then be evaluated.

Nascap-2k is a successor code to NASCAP for geosynchronous orbit (NASCAP/GEO), NASCAP for low-earth orbit (NASCAP/LEO), POLAR, and dynamic plasma analysis code (DynaPAC). NASCAP/GEO has been the standard 3-D tool for the computation of spacecraft charging in tenuous plasmas since 1980. In the following two decades, the fully 3-D computer codes NASCAP/LEO, POLAR, and DynaPAC were developed to address various other spacecraft–plasma interaction issues. Nascap-2k incorporates almost all of the physical and numeric models of these earlier codes. Nascap-2k is available on request to U.S. citizens only; a Web reference with access and other material is http://see.msfc.nasa.gov (SEE Products: Electromagnetic Effects & Spacecraft Charging).

C.3.4 SEE Interactive Spacecraft Charging Handbook

The SEE Interactive Spacecraft Charging Handbook is an interactive spacecraft charging code for the nonexpert. It computes spacecraft surface charging for geosynchronous and auroral zone spacecraft along with internal charging related to the deposition of high-energy (MeV) electrons. Eight assessment modeling tools are included: Geosynchronous Environment, Aurora Environment, Trapped Radiation Environment, Material Properties, Single Material Surface Charging, Multi-Material Surface Charging, Three-Dimensional Surface Charging, and Internal Charging. It can be obtained through the Web site http://see.msfc.nasa.gov (SEE Products: Electromagnetic Effects & Spacecraft Charging). Contact: barbara.m.gardner@saic.com.

C.3.5 Spacecraft Plasma Interaction System (SPIS)

The SPIS software project aims at developing a software toolkit for spacecraft–plasma interactions and spacecraft charging modeling. SPIS is developing a charging code that includes electrical circuit parameters and can model the time behavior of charging and discharge currents. The project was started in December 2002 and has three major objectives:

1. To build the architecture for the SPIS being developed
2. To implement the physical routines of the code
3. To organize and coordinate with the Spacecraft Plasma Interaction Network (SPINE) community

The overall project has been undertaken within the framework of SPINE. The first development phase of the project has been performed by the French Aerospace Lab (ONERA)/Space Environment Department (DESP), Artenum, and University Paris VII under an ESA contract. Further information is available at http://dev.spis.org/projects/spine/home/spis.

REFERENCES

[1] H. B. Garrett, *Guide to Modeling Earth's Trapped Radiation Environment*, AIAA G-083-1999, ISBN 1-56347-349-6, American Institute of Aeronautics and Astronautics, Reston, Virginia, 55 pages, 1999.

[2] D. J. Rodgers, K. A. Hunter, and G. L. Wrenn, "The FLUMIC Electron Environment Model," *8th Spacecraft Charging Technology Conference*, Huntsville, Alabama, October 20–24, 2004.

[3] A. Jursa, Ed., *Handbook of Geophysics and the Space Environment*, Air Force Geophysics Laboratory, U.S. Air Force, National Technical Information Service Document, Accession AD-A167000, 1985. Additional reading for the space environment and interactions with spacecraft. An excellent reference for Earth space plasma environments as well as many other space environments.

[4] J. I. Minow, A. Diekmann, and W. Blackwell, Jr., "Status of the L2 and Lunar Charged Particle Environment Models," *45th AIAA Aerospace Sciences Meeting and Exhibit*, Reno, Nevada, AIAA Paper 2007-0910, 2007.

[5] J. I. Minow, W. C. Blackwell, Jr., L. F. Neergaard, S. W. Evans, D. M. Hardage, and J. K. Owens, "Charged Particle Environment Definition for NGST: L2 Plasma Environment Statistics," *Proceedings of SPIE 4013, UV, Optical, and IR Space Telescopes and Instruments VI*, pp. 942–953, 2000.

[6] J. I. Minow, W. C. Blackwell, Jr., and A. Diekmann, "Plasma Environment and Models for L2," *42nd AIAA Aerospace Sciences Meeting and Exhibit*, Reno, Nevada, AIAA Paper 2004-1079, 2004.

[7] J. I. Minow, L. N. Parker, and R. L. Altstatt, "Radiation and Internal Charging Environments for Thin Dielectrics in Interplanetary Space," *9th Spacecraft Charging Technology Conference*, Tsukuba, Japan, April 2005.

[8] *Space Environment for USAF Space Vehicles*, MIL-STD-1809 (USAF), U.S. Air Force, 69 pages, February 15, 1991.

[9] H. B. Garrett and S. E. DeForest, "Analytical Simulation of the Geosynchronous Plasma Environment", *Planetary and Space Science*, vol. 27, pp. 1101–1109, 1979.

[10] W. R. Nelson, H. Hirayama, and D. W. O. Rogers, *The EGS4 Code System*, SLAC-265, Stanford Linear Accelerator Center, Stanford University, Stanford, California, December 1985.

[11] J. A. Hableib, R. P. Kensick, T. A. Melhorn, G. D. Valdez, S. M. Seltzer, and M. J. Berger, *ITS 3.0: Integrated Tiger Series of Coupled Electron/Photon Monte Carlo Transport Codes*, Radiation Safety Information Computational Center, Oak Ridge, Tennessee, November 1994. Provides additional information on ITS.

[12] A. F. Bielajew, H. Hirayama, W. R. Nelson, and D. W. O. Rogers, "History, Overview and Recent Improvements of EGS4," *Radiation Transport Calculations Using the EGS4 Conference*, Capri, Italy, 1994.

[13] S. Agostinelli, J. Allison, K. Amako, J. Apostolakis, H. Araujo, P. Arce, M. Asai, et al., "Geant4-A Simulation Toolkit", *Nuclear Instruments and Methods in Physics Research*, vol. A, no. 506, pp. 250–303, 2003.

[14] J. Allison, K. Amako, J. Apostolakis, H. Araujo, P. Arce Dubois, M. Asai, G. Barrand, et al., "Geant4 Developments and Applications", *IEEE Transactions on Nuclear Science*, vol. 53, no. 1, pp. 270–278, 2006.

[15] J. F. Briesmeister, Ed., *MCNP-4B: A General Monte Carlo N-Particle Transport Code, Version 4B*, Report LA-12625-M, Los Alamos National Laboratory, New Mexico, March 1997.

[16] T. M. Jordon, *NOVICE: A Radiation Transport/Shielding Code; User's Guide*, Experimental and Mathematical Physics Consultants, Gaithersburg, Maryland, January 2, 1987.

[17] S. M. Seltzer, *SHIELDOSE: A Computer Code for Space-Shielding Radiation Dose Calculations*, NBS Technical Note 1116, National Bureau of Standards (now National Institute of Standards and Technology), U.S. Government Printing Office, Washington, D.C., 1980.

[18] J. Sorensen, D. J. Rodgers, K. A. Ryden, P. M. Latham, G. L. Wrenn, L. Levey, and G. Panabiere, "ESA's Tools for Internal Charging", *IEEE Transactions on Nuclear Science*, vol. 47, no. 3, pp. 491–497, June 2000. A published reference for DICTAT.

[19] S. Hosoda, S. Hatta, T. Muranaka, J. Kim, N. Kurahara, M. Cho, H. Ueda, K. Koga, and T. Goka, "Verification of Multi-Utility Spacecraft Charging Analysis Tool (MUSCAT) via Laboratory Test," *45th AIAA Aerospace Sciences Meeting and Exhibit*, Reno, Nevada, AIAA 2007–278, January 8–11, 2007.

[20] M. J. Mandell, V. A. Davis, B. M. Gardner, I. G. Mikellides, D. L. Cooke, and J. Minor, "Nascap-2k—An Overview", *Transactions on Plasma Science*, vol. 34, no. 5, pp. 2084–2093, 2006.

[21] V. A. Davis, M. J. Mandell, B. M. Gardner, I. G. Mikellides, L. F. Neergaard, D. L. Cooke, and J. Minow, "Validation of Nascap-2k Spacecraft–Environment Interactions Calculations," *8th Spacecraft Charging Technology Conference*, Huntsville, Alabama, in NASA Technical Reports Server, 2004.

APPENDIX D
Internal Charging Analyses

See Appendix G for surface charging analyses.

D.1 THE PHYSICS OF DIELECTRIC CHARGING

As stated earlier, the computations involved in estimating dielectric charging resemble surface charging calculations with the inclusion of space charge. That is, the basic problem is the calculation of the electric field and charge density in a self-consistent fashion over the volume of interest. In other words, Poisson's equation is solved subject to the continuity equation. The relevant formulas are Poisson's equation (in one dimension):

$$\frac{\partial(\varepsilon(x)E(x,t))}{\partial x} = \rho(x,t) \tag{D.1-1}$$

the continuity equation (in one dimension):

$$\frac{\partial\rho(x,t)}{\partial t} = -\frac{\partial(J_c(x,t) + J_R(x,t))}{\partial x} \tag{D.1-2}$$

and Ohm's law (for electrons):

$$J_c(x,t) = \sigma(x,t)E(x,t) \tag{D.1-3}$$

These can be combined to give

$$\frac{\partial(\varepsilon(x)E(x,t))}{\partial t} + \sigma(x,t)E(x,t) = -J_R(x,t) \tag{D.1-4}$$

where
- E = electric field at x for time t
- ρ = charge density at x for time t
- σ = conductivity in $(\Omega \cdot \text{cm})^{-1} = \sigma_0 + \sigma_r$

Guide to Mitigating Spacecraft Charging Effects, First Edition. Henry B. Garrett, Albert C. Whittlesey.

σ_0 = dark conductivity
σ_r = radiation-induced conductivity
$\varepsilon = \varepsilon_0\varepsilon_r$
ε_0 = free-space permittivity = 8.8542×10^{-12} F/m
ε_r = relative dielectric constant
J_R = incident particle flux (current density) where - $\partial J_R/\partial x$ = charge deposition rate at x
J_c = particle flux (current density) due to dark conductivity at x

This equation follows from Poisson's equation and current continuity with the total current, consisting of the incident current J_R (primary and secondary particles) and a conduction current σE. It is solved at a given time t to give the charge variations in x in the dielectric. The results are then stepped forward in time to compute the time-varying charge and electric field.

A simple solution for this equation assuming that σ and J_R are independent of time for a dielectric between two metal plates with an initial imposed field is

$$E = E_0 \exp\left(-\frac{\sigma t}{\varepsilon}\right) + \frac{J_R}{\sigma}\left[1 - \exp\left(-\frac{\sigma t}{\varepsilon}\right)\right] \tag{D.1-5}$$

where E_0 is the electric field imposed at $t = 0$. This is only a crude approximation to reality, as geometrical effects, time variations in the conductivity and incident current, and other effects make numerical solution a necessity. It is, however, useful in understanding the time constants ($\tau = \varepsilon/\sigma$) involved in charging the dielectric—as time increases, the initial field E_0 dies away, tending toward the radiation-induced field given by J_R/σ with a time constant of τ and $\sigma = \sigma_0 + \sigma_r$. Typical values for τ range from about 10 s to 10^3 s for $10^{-16} < \sigma < 10^{-14}$ $(\Omega \cdot \text{m})^{-1}$. Where the dose rate is high (enhancing the radiation conductivity σ_r), the E-field comes to equilibrium rapidly. In lightly irradiated regions, where the time constant is long (the dark conductivity σ_0 dominates), the field takes a long time to reach equilibrium.

The peak electric field (E_{max}) in the irradiated dielectric has been estimated [1] for radiation with a broad energy distribution to be

$$E_{\text{max}} = \frac{A/k}{1 + \sigma/kD} \sim \frac{A}{k} \tag{D.1-6}$$

where
A = 10^{-8} s $\cdot$ V/$\Omega \cdot$ rad $\cdot$ m^2
k = coefficient of radiation induced conductivity (s/m $\cdot\ \Omega \cdot$ rad)
D = average dose rate (rad/s)

The second approximation follows for high flux conditions [1] when the radiation conductivity σ_r can be approximated by

$$\sigma_r \sim kD^{\delta} \tag{D.1-7}$$

where $\sigma_r > \sigma_0$ for high fluxes and $\delta \sim 1$. This equation is in agreement with analytic solutions when they exist and, for some configurations, more complex numerical solutions. Typical values of k are $10^{-16} < k < 10^{-14}$ for polymers [1]. Inserting the range of values for k, E_{max} varies up to 10^6 to 10^8 V/m, respectively, the range where breakdowns are expected.

This simple analysis demonstrates several important concepts. First, by charging a dielectric surface and measuring how long it takes for the charge to bleed off [in the absence of radiation-induced conductivity (RIC)], one can estimate σ_0 from $\sigma_0 = \tau\varepsilon$, where τ is measured by the experiment. In the presence of radiation, the foregoing demonstrates how the charge can be bled off by the RIC σ_r. The equations imply that σ_r is proportional to dose. Ultimately, these equations can be used to estimate whether the potential will build up sufficiently in a dielectric to cause arcing—the key issue of concern here.

D.2 SIMPLE INTERNAL CHARGING ANALYSIS

The following example of a simple and conservative analysis (Table D-1) will be used to estimate the current flux deposited in a dielectric of a spacecraft at GEO. This method of analysis has matched a TIGER internal charging analysis to within 40% or better; it provides a good start to determine if there is a level of concern. If the simple analysis indicates that the flux is close to the design limit, a complete analysis should be used to determine if the criteria is exceeded. In fact, if the simple analysis shows a level of concern, the region in question should probably have its design changed, if possible, or otherwise protected from internal charging. The example determines the flux of electrons in a 10-mil-thick

Table D-1. Simple Charging Example

Electron Flux	Penetration Energy	Exiting Integral Flux
(1) Into 10 mils of aluminum	~250 keV	6×10^6 e/cm$^2 \cdot$ s $\cdot$ Sr
(2) Through 10 mils of Teflon® (equivalent to 7.8 mils of aluminum, total 17.8 mils)	~300 keV	4.5×10^6 e/cm$^2 \cdot$ s $\cdot$ Sr
(3) The net electron flux in the Teflon® is:	$j1 = 6–4.5 \times 10^6$ e/cm$^2 \cdot$ s $\cdot$ Sr $= 1.5 \times 10^6$ e/cm$^2 \cdot$ s $\cdot$ Sr	
(4) Convert to normal incidence flux:	$j2 = \sim j1 \times 3 = 1.5 \times 3 \times 10^6$ e/cm$^2 \cdot$ s $= 4.5 \times 10^6$ e/cm$^2 \cdot$ s	
(5) Convert flux to current in the Teflon®:	$I = 1.602 \times 10^{-19} \times 4.5 \times 10^6 = 0.72$ pA/cm^2	

layer of Teflon® under a 10-mil-thick sheet of aluminum. Figure 2-3 provides mean penetration depth versus energy, and Fig. 2-6 presents fluxes versus energy. Tables 6-1 and 6-2 list material densities.

In this example, the electron charge/flux entering and exiting each layer is calculated; the difference is the electron flux deposited in that layer. For dielectrics, if the deposited current in the layer is > 0.2 pA/cm^2, that is considered as a potential concern, and a more exact analysis should be done. The assumed electron environment is the worst-case GEO environment, as shown in Fig. 2-6. In Fig. 2-3, 10 mils of aluminum require 250-keV energy electrons to penetrate the aluminum and enter the Teflon®. Teflon® density is 78% of aluminum (Table 6-1); therefore, 10 mils of Teflon® is equivalent to 7.8 mils of aluminum. Electrons with greater than 300 keV can penetrate through the 17.8-mil aluminum equivalent and exit the sandwich. Referring to Fig. 2-6, the worst-case flux entering the Teflon® is about 6×10^6 e/cm$^2 \cdot$ s $\cdot$ Sr while the exiting flux is about 4.5×10^6 e/cm$^2 \cdot$ s $\cdot$ Sr, leaving a net flux rate of accumulation of 1.5×10^6 e/cm$^2 \cdot$ s $\cdot$ Sr in the Teflon®. Equivalent normally incident flux is more than the omnidirectional flux. For this simple example covered by 10 mils of aluminum, it is taken to be a factor of 3 times the omnidirectional flux. Converting to current requires multiplying by 1.602×10^{-19} A/e $\cdot$ s. The net (approximate) result is that the charging rate in the 10-mil layer of Teflon® is 0.72 pA/cm^2.

According to Fig. 2-5 and Section 3.2.3.2.2, the charging rate in this Teflon® sample exceeds the safe level of 0.1 pA/cm^2. Therefore, this sample is threatened by occasional discharges. More than 10 mils of aluminum shielding equivalent is required on top of this sample to reduce the charging rate in the Teflon® layer to less than 0.1 pA/cm^2.

Note: The analysis in this section uses deposited flux of 0.1 pA/cm^2 as a criterion rather than incident flux of 0.1 pA/cm^2 as used throughout the rest of this document. This is less conservative than the incident flux criterion. A flux of 0.1 pA/cm^2 in 10 h accumulates 2×10^{14} e/m^2 in 10 h, which will create an electric field of 2×10^6 V/m ($\varepsilon_r = 2$) if all electrons stop in the material in accordance with the criterion used in this paragraph. Assuming an incident flux of 0.1 pA/cm^2 will be more conservative because not all electrons will be stopped in the material. The latter assumption is the better one unless the dielectric strength of the material in question is known to be high, as in this example for Teflon®. See reference [2] which challenges the 10-h accumulation time for highly resistive materials.

D.3 DETAILED ANALYSIS

A proper analysis should be performed using the models and tools listed in Appendixes B and C to determine charge deposition rates (fluxes and fluences). The analysis should determine if sufficient charge exists for breakdown (ESDs). Detailed formulations (e.g., NUMIT and DICTAT) have been developed for determining the development of electric fields in irradiated insulators. In the end, for good insulators at high fluxes, the electric field builds up to and stabilizes at 10^5 and rarely to 10^6 V/cm (10^7 to 10^8 V/m).

The conductivity of the material is a critical parameter to assess the breakdown fields and generally is not known well enough to provide meaningful calculations. For proper answers, one should know the conductivity under irradiation, temperature, and vacuum to perform a meaningful detailed analysis. Even then, predicting pulse amplitudes and rates is only a guess.

As a matter of comparison, a computer code was used to replicate the previous simple example. The results were that the electron flux in the Teflon® was computed to be about 40% of the result from the simple analysis. This shows that for the test case, the simple analysis was conservative by a factor of 2.5. Although shown to be conservative as calculated for the case shown in Section D.2, the simple analysis should always be treated with some suspicion.

Note: TIGER calculations have demonstrated that tantalum reflects some electrons at the surface, and thus the simple calculations above will lead to higher deposited electron fluences than in the actual case (our one example had double the fluence of the TIGER-calculated case). Other physics effects may also be present. Fortunately, this phenomenon does not happen for aluminum.

D.4 SPACECRAFT LEVEL ANALYSIS

A spacecraft level of analysis is used to predict the current density (flux) within the spacecraft interior. It can use radiation analysis tools modified as required to accomplish the task. Conventional radiation analyses inside a spacecraft use transport codes to carry out 3-D tracking of energetic particles through the spacecraft walls to a specific target. The output of these codes is the radiation dose as a function of a detector material (usually Si). Several computer codes that use electron spectra and spacecraft geometry as inputs can also be used to determine internal fluxes or radiation dose at specific sites (Appendix C). This is first done with only the walls and shelves in place. Once the isoflux contours are determined, the flux levels are compared to the critical flux level. If the levels predicted exceed the critical levels, a box-level analysis is conducted. If the flux level inside the box still exceeds the critical flux level, additional shielding should be considered.

The criterion to be used for IESD is that the current (flux) should be less than 0.1 pA/cm^2 for any period of 10 h. If this criterion is satisfied, there should be few problems with internal charging.

D.4.1 Dose-to-Fluence Approximation

To determine an approximate electron flux/fluence from a radiation transport code, a simple equivalence from dose (rad-Si) to electron fluence can be used if the dose has already been calculated or if it is easier to calculate dose. Dose and fluence are related by [3,4]

$$\text{fluence}\,(\text{e/cm}^2) = 2.4 \times 10^7 \times \text{dose}\,(\text{rad-Si}) \qquad \text{(D.4-1)}$$

Although the actual conversion factor varies with energy, this equation is valid for electron energies from about 0.2 to 30 MeV. This is adequate for most internal charging assessments based on typical space environments and can be used for lower energies without loss of "back-of-the-envelope" accuracy already inherent in this method.

As the results from this simple conversion are typically conservative (it predicts greater electron fluence than actually exists), its use would lead to a conservative design and hence greater cost. Coakley [5], for example, says that a 416-krad dose is equivalent to 2×10^{13} e/cm^2 fluence, or fluence (e/cm^2) = $5 \times 10^7 \times$ dose (rad-Si). This is within a factor of 2 of Eq. (D.4-1).

REFERENCES

[1] A. R. Frederickson, D. B. Cotts, J. A. Wall, and F. L. Bouquet, *Spacecraft Dielectric Material Properties and Spacecraft Charging*, AIAA Progress in Astronautics and Aeronautics, Vol. 107, American Institute of Aeronautics and Astronautics, Washington, D.C., 1986. Contains dielectric properties data, especially relating to spacecraft charging. Worth obtaining and reading.

[2] M. Bodeau, "High Energy Electron Climatology That Supports Deep Charging Risk Assessment in GEO," *48th AIAA Aerospace Sciences Meeting*, Orlando, Florida, AIAA 2010-1608, 2010. A fine work with good concepts, explained and illustrated with actual space data and estimates of fluence accumulation versus material resistivity. Bodeau challenges the 0.1-pA/cm^2 and 10-h flux integration guidelines.

[3] E. P. Wenaas, M. J. Treadaway, T. M. Flanagan, C. E. Mallon, and R. Denson, "High-Energy Electron-Induced Discharges in Printed Circuit Boards", *IEEE Transactions on Nuclear Science*, vol. NS-26, no. 6, pp. 5152–5155, 1979.

[4] J. W. Haffner, G. Gigas, J. E. Bell, D. T. Butcher, R. A. Kjar, C. T. Kleiner, and G. C. Messenger, *The Effects of Radiation on the Outer Planets Grand Tour*, SD 71–770, NASA-CR-127065, Jet Propulsion Laboratory, Pasadena, California, 316 pages, November 1971.

[5] P. Coakley, *Assessment of Internal ECEMP with Emphasis for Producing Interim Design Guidelines, JAYCOR Report*, AFWL-TN-86-28, Air Force Weapons Laboratory, Kirtland Air Force Base, New Mexico, June 1987.

APPENDIX E
Test Methods

Tests that can be performed to validate some aspects of charging problems are described conceptually below. The focus here is largely on materials with limited descriptions of component, subsystem, and system tests. Details such as test levels, test conditions, instrumentation ranges, bakeout time, and pass/fail criteria should be considered for any tests. Vacuum bakeout/aging of materials before testing is important because apparent surface properties, especially resistivity, quite often increase with aging in space, as adsorbed water and other conductive contaminants depart because of outgasing.

E.1 ELECTRON-BEAM TESTS

Electron-beam test facilities are to be used to test smaller elements of the spacecraft. This test can be used to determine whether a material sample will arc in a given electron environment and can measure the size of the resulting ESD, if any. Electron-beam tests have the advantage that they are real: The electrons can be accelerated to energies that will penetrate and deposit more or less to the depth desired by the experimenter. They have the disadvantage that the beam is usually monoenergetic rather than a spectrum—the electrons will be deposited initially in a diffuse layer depending on their energy rather than being distributed throughout the exposed material. Usually, the illuminated area is less than 10^3 cm^2 in size. The real area may not be testable, in which case scaling should be applied to the measured results to estimate the real threat. A typical test configuration in a vacuum chamber is shown in Fig. E-1.

The electron source should have both the requisite energy (usually expressed in keV or MeV) and the requisite flux [expressed as a current (pA/cm^2) or flux (e/cm$^2 \cdot$ s)]. (*Note:* 1 pA/cm^2 = 6.242 × 10^6 e/cm$^2 \cdot$ s.) The target material in Fig. E-1 shows a grounded backplate. Some tests may involve a front metal plate, grounded or ungrounded, to simulate the in-flight hardware more closely. In this example, the electrons, after deposition on or in the target material, may leak off to the backplate, or they may remain in the material if its resistivity is high.

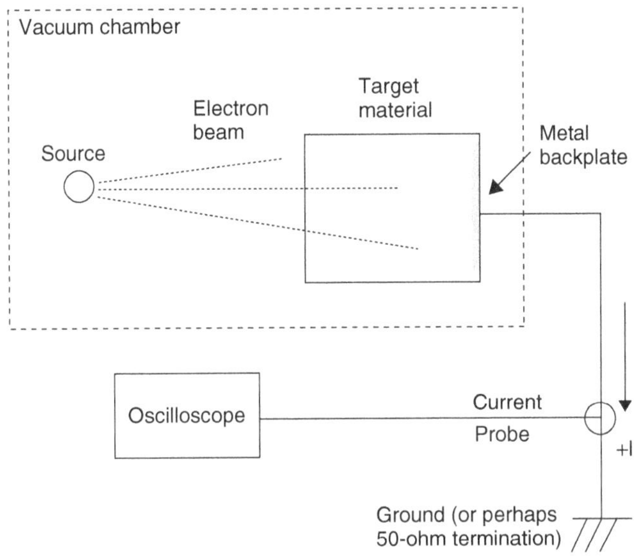

Fig. E-1. Typical electron-beam test facility setup.

If they do not leak off to the backplate (harmlessly), they continue accumulating until the electric field exceeds the dielectric strength of the material and an ESD occurs.

The current probe and oscilloscope are used to determine the current waveform of the ESD from the material. If a simple breakdown between the material and the metal backplate occurs, the current probe can measure the discharge directly. From the waveform, the peak current, the pulse width, and the charge are calculated. If there is a 50-Ω termination, the voltage waveform can be measured and the power and energy in the discharge estimated.

The best way to test a dielectric for IESD is to use an electron beam that penetrates to the middle of the thickness. First, dry the sample in vacuum (drying for a month is best), then irradiate at 1 to 10 nA/cm^2 for several hours and monitor all wires. A sample that does not arc after this test will be excellent in space.

Other diagnostics can be included, including a Rogowski coil to measure electrons blown off the front surface of the material to "space" (the chamber walls) or RF field sensors (EMC antennas and receivers) to measure the spectrum of the radiated noise.

E.2 DIELECTRIC STRENGTH/BREAKDOWN VOLTAGE

This number can be used for ESD analyses to determine the magnitude of the ESD. Usually, the dielectric strength (breakdown voltage) of a (dielectric) material is determined from published tables. If necessary, a test can be performed as shown in Fig. E-2. ASTM D-3755-97, Standard Test Method for Dielectric Breakdown Voltage and Dielectric Strength of Solid Electrical Insulating Materials

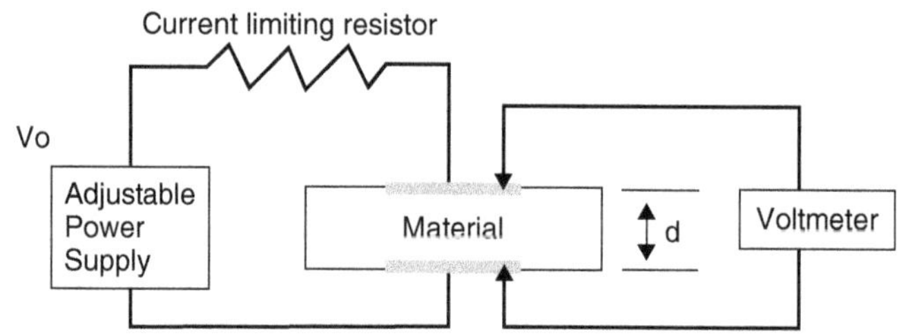

Fig. E-2. Testing for breakdown voltage.

Under Direct-Voltage Stress [1], is a standard test method for breakdown voltage. Normal precautions are to use mechanically sound and clean samples of the material under test. Generally, for any materials involved in internal charging studies, it is appropriate to have a vacuum bakeout to remove the adsorbed water and other contaminants. The test is intended to measure the applied voltage until breakdown. The result is the dielectric strength, which is often reported as V/mil of thickness. The result should also report the tested thickness: V/mil at thickness d.

E.3 RESISTIVITY–CONDUCTIVITY DETERMINATION

Volume conductivity and resistivity are reciprocals of each other. Rho (ρ, $\Omega \cdot$ m) =1/sigma [σ, siemens (S), mho/m, or 1/ $\Omega \cdot$ m]. The volume resistivity of a material is a useful parameter for internal charging assessments. Volume resistivity refers to the bulk resistance of a volume of material. Volume resistivity is determined in terms of the equations supporting Fig. E-3. If the material's volume resistivity is not found in existing tables or the manufacturer's data, it can be measured in one of several ways, as described in the following

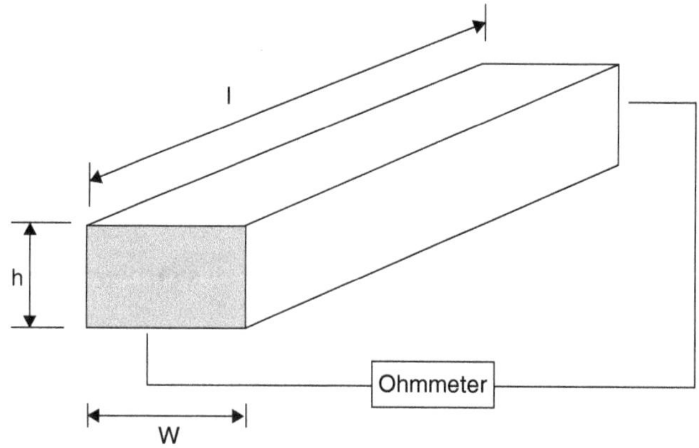

Fig. E-3. Testing for volume resistivity.

paragraphs. ASTM D-257-07, Standard Test Method for DC Resistance or Conductance of Insulating Materials [2], is a standard test method for dc resistance or conductance.

There is another resistivity, surface resistivity, which is applicable to thin layers of material or surface coatings. Surface resistivity, ρ_s (rho–sub s) is the resistance of a flat 2-D square piece of material as measured from one edge to an opposite edge. It may also refer to a surface layer of conductivity on an insulator, which, if the surface has been contaminated by handling or processing, may differ significantly from the bulk resistivity. The resistance of a 2-D surface measured in this manner will be

$$R = \rho_s \frac{l}{w} \tag{E.3-1}$$

where

R = resistance of the sample as measured from end to end (Ω)
ρ_s = surface resistivity (Ω or Ω/square)
l = length of sample, with ground connections at the ends (m)
w = width of sample (m)

For a square sample (length equals width), it can be seen that the resistance from edge to edge will have the same value regardless of the size, so surface resistivity is sometimes called "ohm per square," although the proper unit is simply Ω.

E.4 SIMPLE VOLUME RESISTIVITY MEASUREMENT

Figure E-3 shows the concept of resistivity. The resistance from end to end of the material is as follows:

$$R = \rho \frac{l}{h \times w} \tag{E.4-1}$$

where ρ is the volume resistivity (ohm $\cdot$ m in SI units); sometimes called ρ_v (rho–sub v) and h is the height of the sample (m). Therefore,

$$\rho = R \frac{hw}{l} \tag{E.4-2}$$

Conductivity (S or σ) is the reciprocal of resistivity:

$$S = \frac{1}{\rho} (\text{siemens or } 1/\Omega) \tag{E.4-3}$$

Various difficulties occur when measuring high resistivities, such as higher resistance than can be measured by the ohmmeter, resistivity as a function of voltage stress, resistivity as a function of temperature (more resistive when colder), resistivity modifications related to presence of absorbed moisture, and surface resistivity leakage rather than current flow through the bulk of the material. Test devices such as the Hewlett-Packard Model 4329A high-resistance meter [3], when used in conjunction with a Model 16008A Resistivity Cell [4], can account for some of these problems. That instrument combination can measure very high resistances, has several user-defined test voltages, and has guard rings to prevent surface leakage effects from contaminating the results. The person doing the test should still bake out the test sample to get rid of moisture-caused conductivity. Testing versus temperature is important for cold situations (on the outside of the spacecraft) because resistance is significantly higher at cold space temperatures. For resistances above 10^{11} Ω, moisture bakeout and vacuum tests are appropriate, because moisture adsorption increases conductivity.

Exposure to radiation may increase conductivity (RIC). That is, materials may have more conductivity than measured in a ground environment. The quantitative details of this phenomenon are too involved for this document but in general should not be assumed to be of significant help in the IESD situation.

E.5 ELECTRON-BEAM RESISTIVITY TEST METHOD

This method has the advantage in that it measures the material in a vacuum and in response to an electron beam applying the voltage stress. With a metal front and backplate or plated contacts (or none at all), an electron beam is directed onto the front surface of a flat sample of the material as in Fig. E-4. A non-contacting voltage probe is used to measure the potential on the front surface of the material. A picoammeter then measures the current flowing from the back surface to ground. The volume resistivity is calculated in the manner of Fig. E-3. Shielding is needed to avoid stray electron false data.

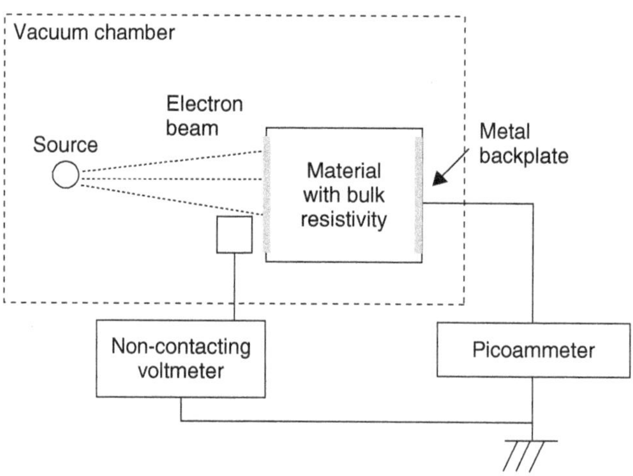

Fig. E-4. Electron-beam test for resistivity.

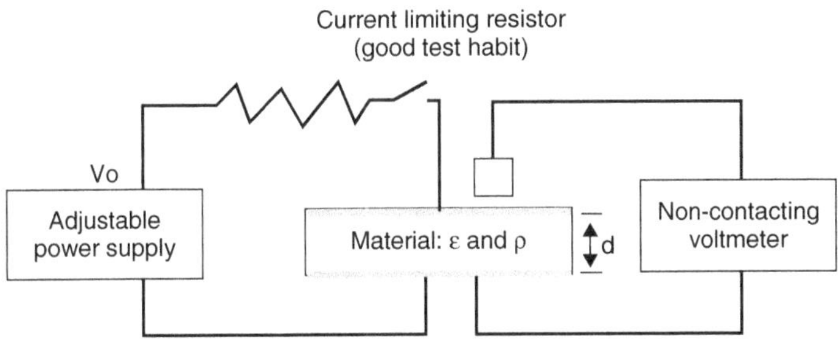

Fig. E-5. Noncontacting voltage-decay resistivity test.

E.6 NONCONTACTING VOLTMETER RESISTIVITY TEST METHOD

This method, illustrated in Fig. E-5, assumes that the resistivity is a constant with respect to applied voltage stress. The method requires plating the upper and lower surfaces of the material being tested to create a capacitor. The capacitance is determined and the capacitor charged. The power supply is disconnected. The voltage decay is monitored as a function of time as measured by a noncontacting voltmeter. The noncontacting voltmeter is necessary because most voltmeters have lower resistance than the test sample and would lead to incorrect measurements. The resistivity is determined by the equations given earlier and by making use of the voltage decay versus time curve, given by the equation

$$V = V_0 e^{-(t/\tau)} \qquad \text{(E.6-1)}$$

where

t = time (s)
$\tau = R \times C$ time constant (s)
R = resistance from top to bottom of the sample (Ω)
C = capacitance of the sample (F)

Problems with this method include the sample preparation (cleanliness, absorbed water, and temperature) and surface leakage around the edge; all should be considered properly. The test could be done in a vacuum chamber to reduce the water absorption contamination of the sample. An electron beam, as shown in Fig. E-4, can be used to charge the sample. The electron beam is then turned off and the voltage decay rate monitored.

Practicalities limit the maximum resistivities measurable with the conventional methods described above. To measure very high resistivities, special techniques are necessary. Dennison et al. [5] describe these methods as used in their laboratory.

E.7 DIELECTRIC CONSTANT, TIME CONSTANT

The dielectric constant, ε, of a material can be determined experimentally, but it almost always can and should be obtained from the manufacturer. From knowledge of permittivity ε and resistivity ρ, the material's relaxation time constant can be determined. One time constant example is the time for a capacitor–resistor combination's voltage to decay to $1/e$ of its full value or about 37% of the original voltage (Fig. E-6). If a rectangular slab of material, as shown in Fig. E-7, has metal electrodes on the top and bottom surfaces, it forms a capacitor, whose value is given by

$$C = \varepsilon\frac{A}{d} \tag{E.7-1}$$

where

ε = permittivity of the material = $\varepsilon_0 \times \varepsilon_r$
ε_0 = permittivity of free space = 8.85×10^{-12} F/m
ε_r = relative dielectric constant of the material, usually between 2 and 4

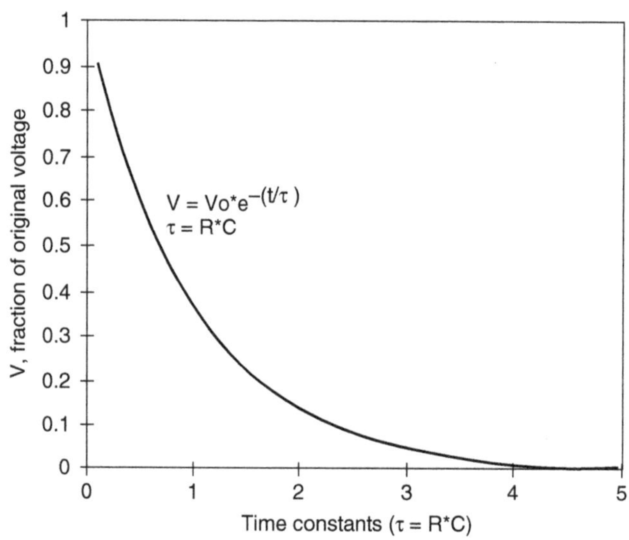

Fig. E-6. *RC* time constants.

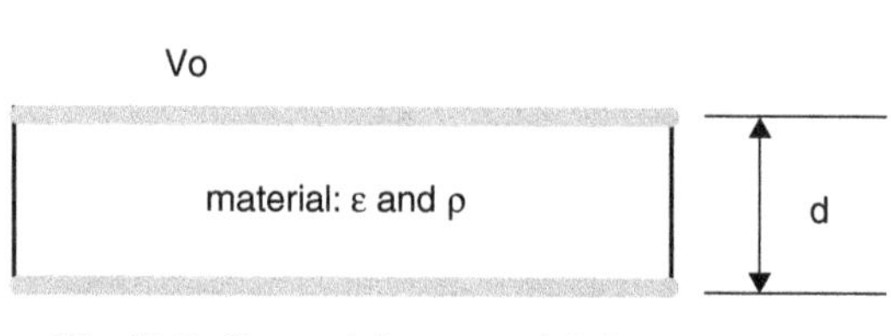

Fig. E-7. Determining material time constant.

A = area of the sample = length × width
d = thickness, top to bottom
R = resistor equivalent to the leakage resistance of the capacitor, computed from the resistivity by standard equations

If the units are the International System of Units (SI), the capacitance will be expressed in farads. Usually, capacitance related to space charging is expressed in picofarads because typical values for space charging are in this range.

The leakage resistance from top to bottom of the same rectangular slab is given by

$$R = \rho \frac{d}{A} \tag{E.7-2}$$

where ρ is the material's volume resistivity (often given in Ω · cm). If the units are consistent, the answer will be in Ω. For the geometry in Fig. E-7, it can be seen that the leakage time constant (τ) is

$$\tau = \rho\varepsilon \tag{E.7-3}$$

At five time constants, there is less than 1% of the original voltage; at 0.01 time constant, the voltage is still 99% of the original. A material time constant of 1 h or less is desirable to leak off detrimental charges before excessive fields cause ESD breakdown in the material [6].

Materials can thus be characterized by their time constants if both the dielectric constant and the resistivity are known. This is a theoretical description. Many high-resistivity materials behave nonlinearly with applied voltage or applied radiation. Thus, these concepts are introductory and approximate. For example, electron-beam tests have found that the discharge time obtained when the beam is turned off (with vacuum maintained) can be hundreds of hours.

E.8 VZAP TEST [MIL-STD-883G, METHOD 3015.7 HUMAN BODY MODEL (HBM)]

A Vzap test is a test of an electronic device's capability to withstand the effects of an electrical transient simulating fabrication handling. It is useful when attempting to decide whether a device can withstand an ESD transient. Figure E-8 shows a typical test configuration (MIL-STD-883G, Method 3015.7 [7]). The parameters are intended to represent the threat from an HBM. The capacitor in this layout (100 ± 10% pF) is charged through $10^6 < \text{R1} < 10^7$ Ω and the power supply is disconnected (switch S1). The capacitor is then discharged (through R2 = 1500 Ω) to the device under test, increasing the voltage until failure. Hardware is classified according to the highest test voltage step that passed without part failure: Class 0 (0 to 249 V), Class 1A (250 to 499 V), Class 1B (500 to 999 V), Class 1C (1000 to 1999 V), Class 2 (2000 to 3999 V), Class 3A (4000 to 7999 V), or Class 3B (>8000 V), depending on its damage threshold.

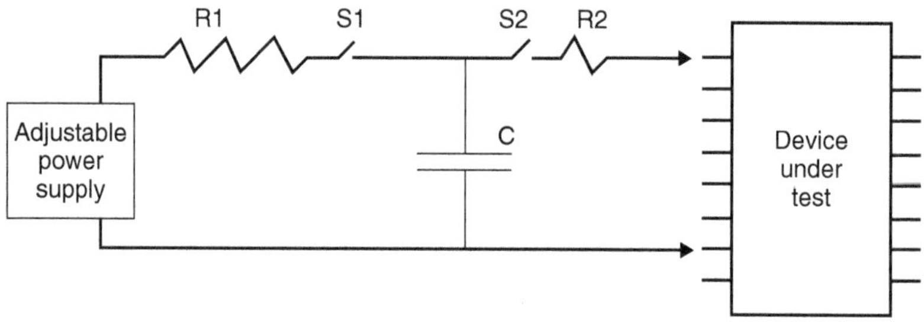

Fig. E-8. Vzap test configuration.

Although providing some idea of the ESD sensitivity of the part, these broad test ranges may not be as precise as desired. This test is mentioned because device sensitivity information may exist from the manufacturer. For actual space discharge events, the value of R2 appears to be in the range 10 to 100 Ω and, more likely, 10 to 50 Ω.

Results obtained by Trigonis [8] for various parts, capacitor sizes, and series resistors (R2) are shown in Fig. E-9, which illustrates how the damage threshold varies with each of the test parameters. Each point represents a different sample for the same part type subjected to a Vzap capacitor discharge at different voltages for various size capacitors. Both polarities are tested and are applied to the weakest pin pairs. The lines plotted show the least energy that damaged any part under any combination of the variables. One feature of the plot is the existence of a minimum-damage voltage threshold for each device. This can be as low as 5 V for some newer devices. The second feature is a constant energy region at low capacitances (not obvious in this figure). The third feature is that the energy appears to go up for the lowest capacitor sizes; this may be an artifact of stray capacitance in the test fixture. It is appropriate to choose the lowest energy as the victim's sensitivity for analyses. It can be seen that for these parts, the weakest component was damaged by 0.5 μJ. Therefore, based on these test results, an ESD needs to deliver at least 0.5 μJ to damage a part. Of course, having data for the actual parts in question is more desirable.

E.9 TRANSIENT SUSCEPTIBILITY TESTS

Transient susceptibility tests are very common in the EMC community. Transient injection is done by inductive or capacitive coupling, as was shown in MIL-STD-462, *Measurement of Electromagnetic Interference Characteristics* [9], for example. The difference between EMC and ESD is the width of the transient pulses: The EMC pulse is typically 10 μs wide, while an ESD pulse is on the order of 10 to 100 ns. A thorough and comprehensive test of a victim device would include varying the pulse width and then determining the voltage and energy threshold of susceptibility. The test should include all pins on the victim device and both polarities of the transient. Testing should include when the input signal is

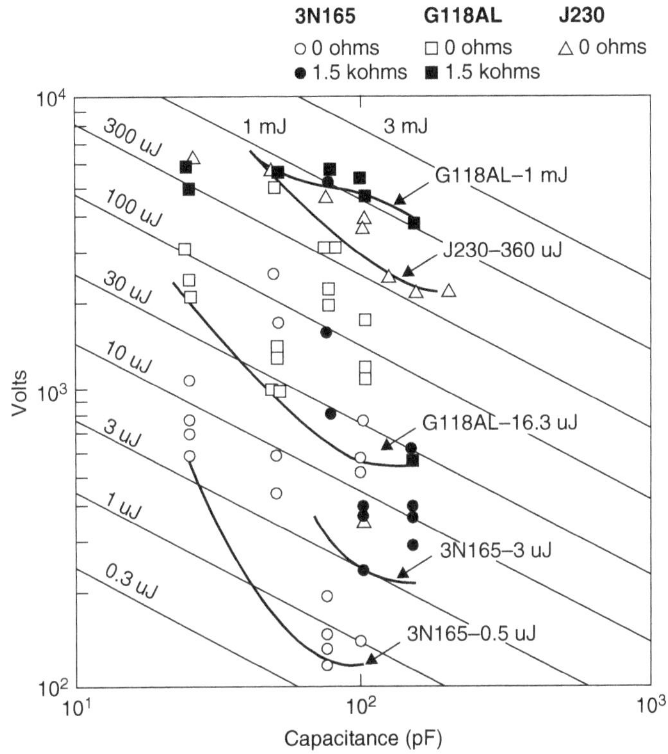

Fig. E-9. Typical results for a Vzap test, showing lines of minimum damage threshold for given parameters. (Based on data collected by Arthur Trigonis [8].) Note: Diagonal lines are for constant energy: $E = 0.5\ C\ V^2$.

in the high state, the low state, and/or transitioning states. Such a comprehensive characterization would involve more work than is usually done, but the analyst should understand that anything less will not be complete.

There are two common sources for generating transient pulses for susceptibility testing. The first is the MIL-STD-1541A [10] pulse source shown in Fig. E-10 and Table E-1. This source provides a capacitive discharge with the amplitude set by the voltage used to charge the capacitor and also the electrode separation gap. The second source is a commercial human body discharge source (Schaeffner supplies one such test device). These sources can be battery operated and also provide a capacitive discharge pulse. The charging voltage is variable so that the amplitude can be controlled. Transients from this source are fast (on the order of 150 ns) and the signal is very clean, as opposed to the MIL-STD-1541A ESD transient source [10]. The state of the art is such that ESD test simulators should be improved to better simulate on-orbit ESD pulses. The reader should search for better sources.

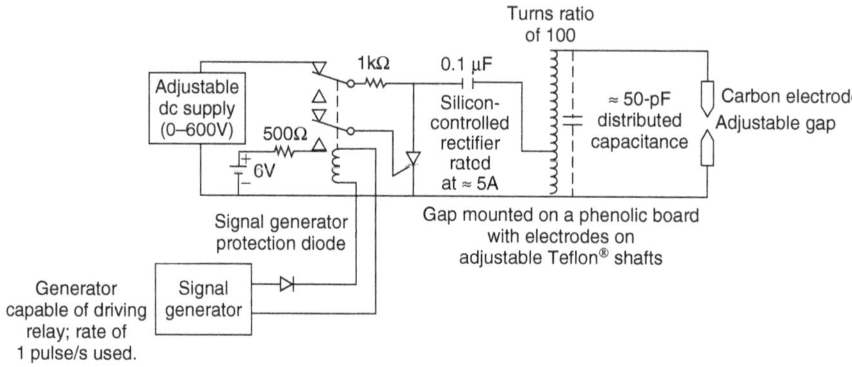

Fig. E-10. MIL-STD-1541A pulse source for transient testing.

Table E-1. Typical Gap Spacing, Voltage, and Energy Levels

Gap (mm)	V_b (kV)	Energy (μJ)
1	1.5	56.5
2.5	3.5	305
5.0	6.0	900
7.5	9.0	2000

E.10 COMPONENT/ASSEMBLY TESTING

Potentially susceptible components and assemblies should be tested for sensitivity to ESD. The component to be tested is to be mounted on a baseplate and functioning. Pulses are to be injected into the component, and the performance of the device is monitored for upsets. The pulses used are to cover the expected range of current amplitudes, voltages, and pulse durations. It is very important that the pulse device be isolated electrically from the component being tested and the monitoring equipment.

E.11 SURFACE CHARGING ESD TEST ENVIRONMENTS

Monoenergetic electron-beam tests have been used to determine approximate surface charging threats of materials.

E.12 SYSTEM INTERNAL ESD TESTING

There is no convenient or cost-effective way to do a system-level internal ESD test.

REFERENCES

[1] Anonymous, *Standard Test Method for Dielectric Breakdown Voltage and Dielectric Strength of Solid Electrical Insulating Materials Under Direct-Voltage Stress*, ASTM D-3755-97, ASTM International, West Conshohocken, Pennsylvania, 2004.

[2] Anonymous, *Standard Test Method for DC Resistance or Conductance of Insulating Materials*, ASTM D-257-07, ASTM International, West Conshohocken, Pennsylvania, 18 pages, 2007.

[3] Anonymous, *Hewlett-Packard Operating and Service Manual for Model 4329A High-Resistance Meter (and Model 16008A Resistivity Cell)*, 1983. With 16008 Resistivity Cell, directly displays high values of resistivity (saves calculation effort).

[4] Anonymous, *Hewlett-Packard Operating Note for Model 16008A Resistivity Cell*, Hewlett-Packard, undated.

[5] J. R. Dennison, J. Brunson, P. Swaminathan, N. W. Green, and A. R. Frederickson, "Methods for High Resistivity Measurements Related to Spacecraft Charging," *IEEE Transactions on Plasma Science*, vol. 34, no. 5, pp. 2191–2203, October 2006. This reference provides a good insight into problems of measuring high resistivities for space usage and proposed test methods appropriate to these needs.

[6] A. R. Frederickson, E. G. Holeman, and E. G. Mullen, "Characteristics of Spontaneous Electrical Discharges of Various Insulators in Space Radiations," *IEEE Transactions on Nuclear Science*, vol. 39, no. 6, pp. 1773–1982, December 1992. This document is a description of the best-known attempt to quantify internal charging effects on orbit, by means of a well-thought-out experiment design. The results were not all that the investigators had hoped, but the data are excellent and very good conclusions can be reached from the data, despite the investigators' concerns.

[7] Anonymous, *Test Method Standard for Microcircuits*, MIL-STD-883G, Method 3015.7 (March 22, 1989), U.S. Department of Defense, 716 pages, February 28, 2006. Method 3015.7 describes Vzap tests for measuring ESD response of electronic parts to the human body model for ESD.

[8] A. Whittlesey, *Example of Semiconductor Damage Thresholds from Capacitor Discharge*, Interoffice Memorandum 5137-11-042, D-69594 (internal document), Jet Propulsion Laboratory, Pasadena, California, May 10, 2011. This useful data set was generated from Arthur Trigonis, "JPL Part Evaluation Report Log 3647," May 1981, a JPL internal working document that has been republished as JPL IOM 5137-11-042 for reference purposes.

[9] Anonymous, *Measurement of Electromagnetic Interference Characteristics*, MIL-STD-462D, U.S. Department of Defense, 189 pages, January 11, 1993.

[10] Anonymous, *Electromagnetic Compatibility Requirements for Space Systems*, MIL-STD-1541A (USAF), U.S. Air Force, 42 pages, December 30, 1987.

APPENDIX F
Voyager SEMCAP Analysis

As an example of a detailed analysis of a spacecraft and the implications of space charging and resulting ESD events, consider the *Voyager* story. To simulate the effects of arc discharges on *Voyager*, tests used a high-voltage-excited spark gap and a flat-plate capacitor with an arc gap to apply arcs on or near the spacecraft. The fields radiated from these sources were approximated in the Specification and Electromagnetic Compatibility Program (SEMCAP) [1], and voltages induced were predicted at key locations using circuit analysis methods. (*Note:* Details are omitted from this appendix because SEMCAP is no longer supported.) Testing then measured induced voltages at those key locations with an oscilloscope. The data measured were compared to the values predicted to give a measure of accuracy of the computational tools. As can be seen, there are enough unknown variables that the results would be expected to differ from reality.

The mean error between the results measured and those predicted was −12 dB (under predicted), and the standard deviation was 20 dB [2]. Assuming these accuracy parameters to be applicable to predicted in-flight responses for *Voyager*, the spacecraft was considered to be immune to arc discharges of less than 20 mV on the basis of the SEMCAP analysis. For research applications, a mean offset of 12 dB and standard deviation of 20 dB sound very large. Despite these estimated accuracies, the use of SEMCAP in this application caused numerous design changes that significantly improved the arc discharge immunity of the *Voyager* spacecraft. Even though *Voyagers* still suffered several arc discharge events in flight, the design changes resulting from SEMCAP (in conjunction with testing) are believed to have enhanced the spacecraft survivability significantly and possibly prevented total failure at Jupiter.

The conclusion here is that even tools providing relatively indeterminate quantitative results can produce results useful for design understanding and possible design changes, as they enforce a systematic approach to evaluating a spacecraft design for ESD.

Guide to Mitigating Spacecraft Charging Effects, First Edition. Henry B. Garrett, Albert C. Whittlesey.

REFERENCES

[1] R. Heidebrecht, *SEMCAP Program Description*, Version 7.4, TRW, Electromagnetic Compatibility Department, Space Vehicles Division, TRW Systems Group, Redondo Beach, California, 1975.

[2] A. C. Whittlesey, "Voyager Electrostatic Discharge Protection Program," *IEEE International Symposium on EMC*, Atlanta, Georgia, pp. 377–383, June 1978.

APPENDIX G
Simple Approximations: Spacecraft Surface Charging Equations

Whereas Appendix D addresses internal charging analyses, in this section we focus on surface charging. The simple approximations discussed in this section are of a worst-case nature. If this analysis indicates differential potentials between noncircuit surface materials of less than 400 V, there should be no spacecraft discharge problems. If potentials predicted on materials exceed 400 V, the Nascap-2k code (Section C.3.3) is to be used.

Although the physics behind the spacecraft charging process is quite complex, the formulation at geosynchronous orbit can be expressed in very simple terms if a Maxwell–Boltzmann distribution is assumed. The fundamental physical process for all spacecraft charging is that of current balance; at equilibrium, all currents sum to zero. The potential at which equilibrium is achieved is the potential difference between the spacecraft and the space plasma ground. In terms of the current [1], the basic equation expressing this current balance for a given surface in an equilibrium situation is

$$I_E(V) - [I_I(V) + I_{SE}(V) + I_{SI}(V) + I_{BSE}(V) + I_{PH}(V) + I_B(V)] = I_T \quad \text{(G-1)}$$

where

V = spacecraft potential
I_E = incident electron current on spacecraft surface
I_I = incident ion current on spacecraft surface
I_{SE} = secondary electron current due to I_E
I_{SI} = secondary electron current due to I_I
I_{BSE} = backscattered electrons due to I_E
I_{PH} = photoelectron current
I_B = active current sources such as charged particle beams or ion thrusters
I_T = total current to spacecraft (at equilibrium, $I_T = 0$)

Guide to Mitigating Spacecraft Charging Effects, First Edition. Henry B. Garrett, Albert C. Whittlesey.

For a spherical body and a Maxwell–Boltzmann distribution, the first-order current densities (the current divided by the area over which the current is collected) can be calculated [1] using the following equations (appropriate for small conducting sphere at GEO):
Electrons:

$$J_E = \begin{cases} J_{E0} \exp\left(\dfrac{qV}{kT_E}\right) & V < 0 \text{ repelled} \quad \text{(G-2)} \\ J_{E0}\left(1 + \dfrac{qV}{kT_E}\right) & V > 0 \text{ attracted} \quad \text{(G-3)} \end{cases}$$

Ions:

$$J_I = \begin{cases} J_{I0} \exp\left(\dfrac{-qV}{kT_I}\right) & V > 0 \text{ repelled} \quad \text{(G-4)} \\ J_{I0}\left(1 - \dfrac{qV}{kT_I}\right) & V < 0 \text{ attracted} \quad \text{(G-5)} \end{cases}$$

where

$$J_{E0} = \frac{qN_E}{2}\left(\frac{2kT_E}{\pi m_E}\right)^{1/2} \quad \text{(G-6)}$$

$$J_{I0} = \frac{qN_I}{2}\left(\frac{2kT_I}{\pi m_I}\right)^{1/2} \quad \text{(G-7)}$$

where

N_E = density of electrons
N_I = density of ions
m_E = mass of electrons
m_I = mass of ions
q = magnitude of the electronic charge
T_E = temperature of electrons
T_I = temperature of ions

Given these expressions and parameterizing the secondary and backscatter emissions, Eq. (G-1) can be reduced to an analytic expression in terms of the potential at a point. This model, called an *analytic probe model*, can be stated as follows:

$$\begin{aligned} &A_E J_{E0}[1 - \mathrm{SE}(V, T_E, N_E) - \mathrm{BSE}(V, T_E, N_E)] \exp\left(\frac{qV}{kT_E}\right) \\ &- A_I J_{I0}[1 + \mathrm{SI}(V, T_I, N_I)]\left(1 - \frac{qV}{kT_I}\right) \\ &- A_{\mathrm{PH}} J_{\mathrm{PH}_0} f(X_m) = I_T = 0 \qquad V < 0 \qquad \text{(G-8)} \end{aligned}$$

where

A_E = electron collection area
J_{E0} = ambient electron current density
A_I = ion collection area
J_{I0} = ambient ion current density
A_{PH} = photoelectron emission area
J_{PH_0} = saturation photoelectron flux
BSE,SE,SI = parameterization functions for secondary emission related to backscatter, electrons, and ions
$f(X_m)$ = attenuated solar flux as a function of altitude X_m of the center of the Sun above the surface of the Earth as seen by spacecraft (%)

This equation is appropriate for a small (< 10 m), uniformly conducting spacecraft at geosynchronous orbit in the absence of magnetic field effects. To solve the equation, V is varied until $I_T = 0$. Typical values for aluminum of SI, SE, and BSE are 3, 0.4, and 0.2, respectively. For geosynchronous orbit, J_E/J_I is about 30 during a geomagnetic storm.

As discussed regarding Eq. (2.3-6), when the spacecraft is in eclipse (and ignoring secondary and backscattered terms), a simple proportionality between the satellite potential and the currents and temperature can be derived from Eq. (G-8):

$$V \sim \frac{-T_E}{q} \times \ln \frac{J_E}{J_I} \tag{G-9}$$

where T_E is in electron volts. That is, to rough order in eclipse, the spacecraft potential is directly proportional to the plasma temperature expressed in electron volts and the natural log of the ratio of the electron and ion currents. Note, however, that secondary currents play a crucial role in actual calculations, and T_E must be greater than some critical value [2–5], usually on the order of 1000 eV, before charging will occur because secondary electron production can exceed the ambient current for low enough T_E. Also, $\ln(J_E/J_I)$ often varies much more rapidly and by larger factors than T_E so that charging has often been found to be more related to changes in $\ln(J_E/J_I)$ than T_E [6].

REFERENCES

[1] H. B. Garrett, "The Charging of Spacecraft Surfaces", *Reviews of Geophysics and Space Physics*, vol. 19, no. 4, pp. 577–616, November 1981. A nice summary paper, with numerical examples and many illustrations. This and Whipple (1981) [7] are two definitive papers on the subject, each covering slightly different aspects.

[2] H. B. Garrett and S. E. DeForest, "Analytical Simulation of the Geosynchronous Plasma Environment", *Planetary and Space Science*, vol. 27, pp. 1101–1109, 1979.

[3] R. C. Olsen, "A Threshold Effect for Spacecraft Charging", *Journal of Geophysical Research*, vol. 88, pp. 493–499, January 1, 1983.

[4] S. T. Lai and D. J. Della-Rose, "Spacecraft Charging at Geosynchronous Altitudes: New Evidence for Existence of Critical Temperature", *Journal of Spacecraft and Rockets*, vol. 38, pp. 922–928, 2001.

[5] V. A. Davis, M.J. Mandell, and M. F. Thomsen, "Representation of the Measured Geosynchronous Plasma Environment in Spacecraft Charging Calculations", *Journal of Geophysical Research*, vol. 113, no. A10204, doi:10.1029/2008JA013116, 2008.

[6] H. B. Garrett, D. C. Schwank, P. R. Higbie, and D. N. Baker, "Comparison Between the 30–80keV Electron Channels on ATS-6 and 1976-059A During Conjunction and Application to Spacecraft Charging Prediction", *Journal of Geophysical Research*, vol. 85, no. A3, pp. 1155–1162, 1980.

[7] E. C. Whipple, "Potentials of Surfaces in Space", *Reports on Progress in Physics*, vol. 44, pp. 1197–1250, 1981. Nice summary paper. Emphasis is on total charging and not internal charging, but good for physics background. This and Garrett [1] are the two definitive papers on that subject, each covering slightly different aspects.

APPENDIX H
Derivation of Rule Limiting Open-Circuit Board Area

This appendix contains equations that describe the rationale for the internal charging design guideline in Section 3.2.3.2.6, Fill Circuit Board Material with Grounded/Referenced Metal, which limits open volumes on a standard circuit board to less than 0.3 cm^2. The assumptions and resulting design guideline presented in this appendix have not been validated by test.

This derivation has been approached as a volumetric equation; that is, the threat is developed on the basis that when a surface area is dielectric, the circuit board under that dielectric is also a dielectric through to the bottom with no ground planes or traces to interrupt the storage of undesired energy in that volume. If there were a ground plane at some depth, these calculations estimate greater storage of energy than actually would be present.

The new rule additionally allows for the presence of ground planes, which reduces the level of concern. The energy of a capacitor of area A and discharge voltage V is

$$E = \tfrac{1}{2}CV^2 \tag{H-1}$$

where

E = energy (J)
C = capacitance (F)
V = voltage (V)

This capacitor is based on FR4 circuit board material with a relative dielectric constant of 4.7 and a 1 cm × 1 cm × ~2 mm (80 mil) thick patch with a 2000-V discharge voltage. The capacitor contains about 2×10^{10} electrons. This quantity of electrons per square centimeter is the amount believed to be critical for internal discharges. The resulting calculated stored energy is about 4 μJ. The design rule is based on protection of a victim with an assumed 1 μJ damage sensitivity, so there should be a limit of approximately 0.3 cm^2 for the area of an empty circuit

Guide to Mitigating Spacecraft Charging Effects, First Edition. Henry B. Garrett, Albert C. Whittlesey.

board region. Extending the equation:

$$C = \varepsilon_0 \varepsilon_r \frac{A}{t} \tag{H-2}$$

where $\varepsilon_0 \varepsilon_r$ is the permittivity of the capacitance material and t is the thickness of the capacitor. If the potential of the discharge voltage is adjusted to be proportional to the thickness ($V = k \times t$) and the results of the equations are combined, we have

$$E = 0.5 \left(\varepsilon_0 \varepsilon_r \frac{A}{t} \right) (k \times t)^2 \tag{H-3}$$

or

$$E = 0.5(\varepsilon_0 \varepsilon_r A) k^2 t \tag{H-4}$$

where k is the dielectric strength (V/m, for example) for the material in question.

The number of electrons implicit in this equation is the same, but the available energy to damage components is proportional to the thickness. If a ground plane (or power plane) is 8 mils below the dielectric surface, the stored energy will be less than a ground plane at 20 mils depth, in proportion to the dielectric thickness, which reduces the level of concern. The ground (or power) plane provides a nearby conductive medium to leak off charge during the charging process. During the discharge process, it provides a nearby location for the discharge to strike and is a much more robust victim than an IC.

A clarification of the rule is that it was based on an assumption that the material in question is approximately square. If it is a long, thin area, it is more difficult to concentrate the ESD energy in one pulse. Therefore, the applicable aspect ratio is 3 : 1. That is, the rule will permit a long patch of dielectric if one dimension is less than 0.3 cm (3 mm). This effect is shown in Fig. H-1, which

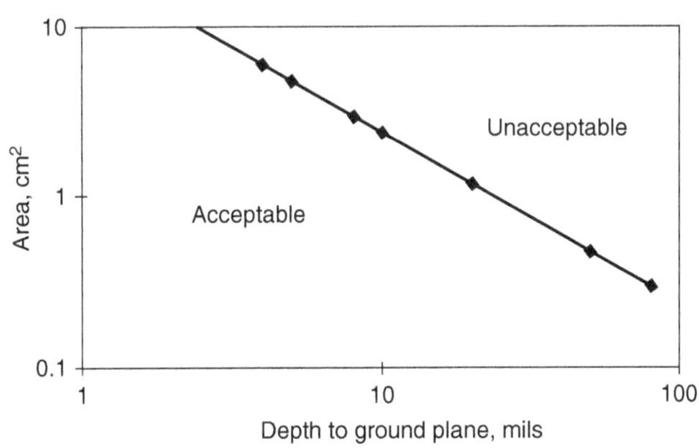

Fig. H-1. Permissible open area of 80-mil-thick FR4 circuit board material versus depth to a ground plane or power plane (preferred) or other circuit traces.

proposes a new rule for exposed dielectric areas on circuit boards. The design rule assumes a standard FR4 circuit board material of 80 mil (~2 mm) thickness.

An experiment was performed to determine energy transfer from an area of circuit board metal to a victim wire [1]. The energy transferred from a charged metal area of 1 cm^2 at breakdown to a nearby trace on the circuit board into a 50-Ω load was about 0.5 μJ, roughly one-tenth the amount calculated above for energy stored in a dielectric volume of the same 1-cm^2 surface area on the 80-mil-thick dielectric. This might indicate a 10% energy transfer efficiency. It provides a rough validation of the analytic results derived above.

REFERENCE

[1] P. L. Leung, G. H. Plamp, and P. A. Robinson, Jr., "Galileo Internal Electrostatic Discharge Program," *Spacecraft Environmental Interactions Technology 1983*, Colorado Springs, Colorado, October 4–6, NASA CP-2359/AFGL-TR-85-0018, National Aeronautics and Space Admin-istration, pp. 423–435, 1983. This paper, documented in the 1985 publication and presented at the *4th Spacecraft Charging Technology Conference*, describes a very neat and clear test that measures the effect of line lengths and circuit board metal areas in the resultant ESD amplitude. It also measures the amplitude of ESD transients from electron beam charging on 50-Ω loads from various conductors.

APPENDIX I
Expanded Worst-Case Geosynchronous Earth Environments Descriptions

The worst-case geosynchronous environment descriptions in Table I-1 are actually several measured environments that caused large spacecraft charging events. They come from documented measured data as referenced and are presented in a form that can be used in a double Maxwellian fit in a charging code such as Nascap-2k. *Note:* In Table I-1, Tav, Trms, and the ATS-6 two-Maxwellian parameters are averaged over all angles. The SCATHA two-Maxwellian parameters are for fluxes parallel and perpendicular to the magnetic field.

There are three environments (one from the Applied Technology Satellite 6 [ATS-6] and two from the Spacecraft Charging at High Altitudes Satellite [SCATHA]), each of which caused large potentials on the respective satellites. Thus, persons wishing to evaluate a satellite design more carefully could make three runs using all three environmental descriptions. In this manner, the design of a satellite could be checked more thoroughly for charging effects in various worst-case environments.

Other attempts at worst-case Earth geosynchronous environments are presented in Table 2-1 (Section 2.2.2, using a single Maxwellian representation of the environment) and in Tables B-1 and B-2 (Section B.2.1). Section B.2.1 suggests a third approach: Start with the average of the appropriate parameter and then increase it by one or more standard deviations, depending on the analyst's opinions.

Guide to Mitigating Spacecraft Charging Effects, First Edition. Henry B. Garrett, Albert C. Whittlesey.

Table I-1. Worst-Case Geosynchronous Environments

Parameter	Deutsch [1] ATS-6	Mullen [2] SCATHA	Mullen [3] SCATHA
Electrons			
Number density (ND) (cm^{-3})	1.22	0.9	3
Current density (J) (nA/cm^2)	0.41	0.187	0.501
Energy density (ED) (eV/cm^3)	2.93×10^4	9.60×10^3	2.40×10^4
Energy flux (EF) ($eV/cm^2 \cdot s \cdot Sr$)	2.64×10^{13}	6.68×10^{12}	1.51×10^{13}
Number density for population 1 (N_1, cm^{-3})	0		
Parallel	—	0.2	1.0
Perpendicular	—	0.2	0.8
Temperature for population 1 (T_1, keV)	0		
Parallel	—	0.4	0.6
Perpendicular	—	0.4	0.6
Number density for population 2 (N_2, cm^{-3})	1.22		
Parallel	—	0.6	1.4
Perpendicular	—	2.3	1.9
Temperature for population 2 (T_2, keV)	16.0		
Parallel	—	24.0	25.1
Perpendicular	—	24.8	26.1
Electron average temperature (T_{av}) (keV)	16	7.7	5.33
Electron root-mean-square temperature (T_{rms}) (keV)	16.1	9	7.33
Ions (Protons)			
Number density (ND) (cm^{-3})	0.245	2.3	3
Current density (J) (nA/cm^2)	0.00252	0.00795	0.0159
Energy density (ED) (eV/cm^3)	1.04×10^4	1.90×10^4	3.70×10^4
Energy flux (EF) ($eV/cm^2 \cdot s \cdot Sr$)	2.98×10^{11}	3.42×10^{11}	7.48×10^{11}
Number density for population 1, (N_1, cm^{-3})	0.00882		
Parallel	—	1.6	1.1
Perpendicular	—	1.1	0.9
Temperature for population 1 (T_1, keV)	0.111		
Parallel	—	0.3	0.4
Perpendicular	—	0.3	0.3
Number density for population 2 (N_2, cm^{-3})	0.236		
Parallel	—	0.6	1.7
Perpendicular	—	1.3	1.6
Temperature for population 2 (T_2, keV)	29.5		
Parallel	—	26	24.7
Perpendicular	—	28.2	25.6
Ion average temperature (T_{av}) (keV)	28.4	5.5	8.22
Ion root-mean-square temperature (T_{rms}) (keV)	29.5	12	11.8

REFERENCES

[1] M.-J. Deutsch, "Worst Case Earth Charging Environment," *Journal of Spacecraft and Rockets*, vol. 19, no. 5, pp. 473–477, 1982.

[2] E. G. Mullen, D. A. Hardy, H. B. Garrett, and E. C. Whipple, "P78-2 SCATHA Environmental Data Atlas," *Spacecraft Charging Technology 1980*, NASA CP 2182/AFGL-TR-81-0270, National Aeronautics and Space Administration, pp. 802–813, 1981.

[3] E. G. Mullen, M. S. Gussenhoven, and H. B. Garrett, *A "Worst Case" Spacecraft Environment as Observed by SCATHA on 24 April 1979*, AFGL-TR-81-0231, Air Force Geophysics Laboratory, Hanscom Air Force Base, Massachusetts, 1981.

APPENDIX J
Key Spacecraft Charging Documents

This appendix is a sampling of the many possible information sources relevant to this field. It is heavily colored by the principal author's knowledge, experience, and prejudices; and it has omitted many worthy references to keep it to a manageable size. The curious reader may dig deeper by following references in these documents. Additional, more specific sources are referenced at the end of each chapter or appendix. The various charging conference records themselves contain a wealth of technical papers.

J.1 U.S. GOVERNMENT DOCUMENTS

J.1.1 DoD

AFGL-TR-77-0288	H. B. Garrett, *Modeling of the Geosynchronous Orbit Plasma Environment*, Part 1, 1978.
AFGL-TR-78-0304	H. B. Garrett, E. G. Ziemba, and S. E. Deforest, *Modeling of the Geosynchronous Plasma Environment*, Part 2, *ATS-5 and ATS-6 Statistical Atlas I*, 1978.
AFGL-TR-79-0015	H. B. Garrett, R. E. McInerney, S. E. Deforest, and B. Johnson, *Modeling of the Geosynchronous Orbit Plasma Environment*, Part *3, ATS-5 and ATS-6 Pictorial Data Atlas*, 1979.
AFRL-VS-TR-20001578	*6th Spacecraft Charging Technology Conference*, October 26–29, 1998, Air Force Research Laboratory, Hanscom Air Force Base, Massachusetts, D. L. Cooke and S. T. Lai, compilers. This conference is documented on one or more CDs, one of which is contained in SEE Publication SEE/TP-2005-600 (J. Minor, compiler, NASA MSFC). The CD contains

Guide to Mitigating Spacecraft Charging Effects, First Edition. Henry B. Garrett, Albert C. Whittlesey.

	photo images of electronic files for the 1st through the 8th Spacecraft Charging Conferences.
AFWAL-TR-88-4143, Vol. II	W. G. Dunbar, *Design Guide: Designing and Building High Voltage Power Supplies*, Materials Laboratory, Air Force Wright Aeronautical Laboratories, Patterson Air Force Base, Ohio, August 1988. Contains good design ideas.
MIL-STD-461	*Requirements for the Control of Electro-magnetic Interference Characteristics of Subsystems and Equipment*. Various versions; version F is the latest as of 2007. Generally, a good EMC design will be helpful at mitigating space charging and ESD effects.
MIL-STD-462	*Measurement of Electromagnetic Interference Characteristics*, July 31, 1967.
MIL-STD-883 G	*Test Method Standard for Microcircuits, Method 3015.7, Electrostatic Discharge Sensitivity Classification*, March 22, 1989. This describes Vzap tests for measuring ESD response of electronic parts to the human body model for ESD.
MIL-STD-1541A	*Electromagnetic Compatibility Requirements for Space Systems*. Appendix E.9 of this handbook has a "Schematic Diagram of Arc Source" as copied from MIL-STD-1541A (December 30, 1987).
MIL-STD-1686	*Electrostatic Discharge Control Program for Protection of Electrical and Electronic Parts, Assemblies, and Equipment (Excluding Electrically Initiated Explosive Devices)*, October 25, 1995.
MIL-STD-1809	*Space Environment for USAF Space Vehicles*, February 19, 1991. This includes some electron spectra that can be used in the electron transport codes. It has good information that supplements Earth environmental information in this version of NASA-HDBK-4002A.
PL-TR-93-2027(I)	*Proceedings of the Spacecraft Charging Technology Conference*, 1989, Volume I., R. C. Olsen, ed., October 31–November 3, 1989, Naval Postgraduate School, Monterey, California. More detailed explanations of the space environment and its interactions with spacecraft.

J.1.2 NASA

NASA-CP-2004-213091	*Spacecraft Charging Technology Conference*, J. L. Minor, compiler, October 20–24, 2003, Marshall Space Flight Center, Huntsville, Alabama. More detailed explanations of the space environment and its interactions with spacecraft.
NASA-CP-2071	*Spacecraft Charging Technology 1978* (also AFGL-TR-79-0082), October 31–November 2, 1978. U.S. Air Force Academy, Colorado Springs, Colorado. More detailed explanations of the space environment and its interactions with spacecraft.
NASA-CP-2182	*Spacecraft Charging Technology 1980* (also AFGL-TR-81-0270), November 12–14, 1980, U.S. Air Force Academy, Colorado Springs, Colorado. More detailed explanations of the space environment and its interactions with spacecraft.
NASA-CP-2359	*Spacecraft Environmental Interactions Technology 1983* (also AFGL-TR-85-0018), October 4–6, 1983, U.S. Air Force Academy, Colorado Springs, Colorado. More detailed explanations of the space environment and its interactions with spacecraft.
NASA-HDBK-4001	*Electrical Grounding Architecture for Unmanned Spacecraft*, February 17, 1998. This is a handy general document. Notice that the grounding diagrams show that the circuit grounds exit the boxes and apparently connect to a remote ground; this is a schematic and not a physical diagram. The grounds should be contained within the box for the EMC reason that it should not act as a radiator (antenna) of noise into or out of the box.
NASA-HDBK-4002	*Avoiding Problems Caused by Spacecraft On-Orbit Internal Charging Effects*, National Aeronautics and Space Administration, Washington, D.C., February 17, 1999. One of the two base documents for NASA-HDBK-4002A along with NASA TP-2361.
NASA-HDBK-4002A	*Mitigating In-Space Charging Effects: A Guideline*, National Aeronautics and Space Administration, Washington, D.C., March 3, 2011.

NASA-HDBK-4006	*Low Earth Orbit Spacecraft Charging Design Handbook*, June 3, 2007. This document is written by two of NASA's senior researchers in spacecraft charging, recognized experts on charging (and discharging) of solar arrays in space plasmas. It is a fine reference to have on your bookshelf for spacecraft charging. See also NASA-STD-4005.
NASA-RP-1354	J. L. Herr and M. B. McCollum, *Spacecraft Environments Interactions: Protecting Against the Effects of Spacecraft Charging*, November 1994.
NASA-RP-1375	R. D. Leach and M. B. Alexander, *Failures and Anomalies Attributed to Space Charging*, August 1995.
NASA-STD-4005	*Low Earth Orbit Spacecraft Charging Design Standard*, June 3, 2007. The NASA standard for LEO charging, it gives mitigation techniques for LEO, some of which are also applicable to GEO and polar environments. See also NASA-HDBK-4006.
NASA TMX-73537	*Proceedings of the Spacecraft Charging Technology Conference* (also AFGL-TR-77-0051), C. P Pike and R. R. Lovell, Eds., October 27–29, 1976, U.S. Air Force Academy, Colorado Springs, Colorado, 1977. More detailed explanations of the space environment and its interactions with spacecraft.
NASA/TP-2003-212287	D. C. Ferguson and G. B. Hillard, *Low Earth Orbit Spacecraft Charging Design Guidelines*, February 2003. See section 3.2.4.2 for added information.
NASA TP-2361	*Design Guidelines for Assessing and Controlling Spacecraft Charging Effects*, 1984. One of the two base documents for NASA-HDBK-4002A. Listed as historical reference; some of the deleted sections can provide more background information and illustrations. Its Section 2.3 describes charge loss in a discharge. Its Section 3.1.2.3 describes retarding potentials on large portions of dielectrics.

J.2 NON-U.S. GOVERNMENT DOCUMENTS

J.2.1 American Society for Testing and Materials (ASTM)

ASTM D-257-61	*Standard Test Methods for DC Resistance or Conductance of Insulating Materials*, 1961. Uses test methods appropriate for normally dielectric materials. For measurement of highly resistive materials often used for space charging applications, special measurement methods should be used. (See ASTM D-257-91.)
ASTM D-257-91	*Standard Test Method for DC Resistance or Conductance of Insulating Materials*, 1991. Good for measuring high values of resistance.
ASTM D-3755	*Standard Test Method for Dielectric Breakdown Voltage and Dielectric Strength of Solid Electrical Insulating Materials Under Direct-Voltage Stress*, March 10, 1997.

J.2.2 European Cooperation for Space Standardization (ECSS)/European Handbooks

These documents, including unpublished drafts, are available via the ESA-ESTEC Web site, http://www.ecss.nl. Accessing documents requires registration on the site, but the only requirements for a user are to supply an ID and e-mail address to:

ECSS Secretariet
ESA-ESTEC
Noordwijk, The Netherlands

ECSS-20-06	*Spacecraft Charging-Environment-Induced Effects on the Electrostatic Behaviour of Space Systems*. Unpublished draft that should be published because of its useful content.
ECSS-E-ST-20-06C	*Space Engineering, Spacecraft Charging Standard*, July 31, 2008. This is intended to be a set of design rules but is far more than that. It contains the background physics and provides a wealth of space charging information, both scientifically and practically oriented. The standard is a very good educational reference. Sometimes, however, it is not explicit in that it may provide two or more answers to the same question (e.g., what environment to use).

J.2.3 European Space Research and Technology Centre

SP-476 *7th Spacecraft Charging Technology Conference 2001: A Spacecraft Charging Odyssey*, April 23–27, 2001, Noordwijk, The Netherlands.

J.2.4 Japanese Aerospace Exploration Agency (JAXA)

SP-05-001E *9th Spacecraft Charging Technology Conference*, T. Goka, compiler, April 4–8, 2005, Epochal Tsukuba, Tsukuba, Japan.

J.2.5 Other

QinetiQ/KI/SPACE/HB042617	D. J. Rodgers, *Spacecraft Plasma Interaction Guidelines and Handbook*, June 30, 2004. Produced for ESA/ESTEC by QinetiQ Ltd, Farnborough, Hampshire, England. Another good reference for persons wishing further background on the subject. Draft accessed at http://www.space.qinetiq.com/spigh/Technical%20note%201.pdf on April 14, 2011.
SD 71–770	*The Effects of Radiation on the Outer Planets Grand Tour*, November 1971 (also NASA-CR-127065, National Aeronautics and Space Administration). Prepared for the Jet Propulsion Laboratory by Space Division, North American Rockwell. This old document is available on the NASA Technical Reports Server http://ntrs.nasa.gov/search.jsp.

APPENDIX K
List of Figures and Tables

FIGURES

Guide to Mitigating Spacecraft Charging Effects, First Edition. Henry B. Garrett, Albert C. Whittlesey.

TABLES

INDEX

Page numbers in *italics* refer to Figures; those in **bold** to Tables

www.ingramcontent.com/pod-product-compliance
Lightning Source LLC
LaVergne TN
LVHW011803200926
848484LV00003B/766
* 9 7 8 1 1 1 8 1 8 6 4 5 9 *